Wave Motion and Applied Science

Wave Motion and Applied Science

Editor by

Jae Lord Dexter C. Filipinas

Delve
PUBLISHING

www.delvepublishing.com

Wave Motion and Applied Science

Jae Lord Dexter C. Filipinas

Delve Publishing
2010 Winston Park Drive,
2nd Floor
Oakville, ON L6H 5R7
Canada
www.delvepublishing.com
Tel: 001-289-291-7705
001-905-616-2116
Fax: 001-289-291-7601
Email: orders@arclereducation.com

ISBN: 978-1-77361-064-1 (Hardcover)

ABOUT THE EDITOR

Jae Lord Dexter C. Filipinas

Jae Filipinas is Assistant Professor in Department of Physics at College of Natural Sciences and Mathematics, Mindanao State University, Philippines. He completed his M.S. Physics from University of Philippines, Diliman , Quezon City, Philippines. He is an avid researcher and has published many proceedings

TABLE OF CONTENTS

PREFACE

Waves have been a significant part of the development of telecommunications on Earth. Wave motion is a propagation of disturbances, which deviates from a state of rest or equilibrium, from place to place in a consistent and arranged way. The most common or familiar are surface waves on water, but sound and light both travel as wavelike disturbances, and the movement of all subatomic particles that displays wavelike characteristics. The study of waves adheres various forms that the topic of central importance in all physical sciences and engineering. Modern applications such as telephone, mobile phone, telecommunications and wireless technologies arise through the effort of studies confined to wave motion and applied science.

The unifying theme of this book is wave motion; its many forms and the changes it undergoes. This book discusses the significance of wave's motion and the applications it brought to the society. The content is divided into the following chapters:

Chapter 1, Wave Motion, discusses the different forms of waves and their propagation. The units and measurement to be applied in the calculations of different kinds of waves, it may either be one dimensional waves or two dimensional waves. In this chapter, most theoretical concepts about waves are presented.

Chapter 2, Light Wave, starts with the definition and properties of light, which is from the big family of Electromagnetic waves. Human eyes are very sensitive to this form of electromagnetic waves. This is the only EM which is visible to the naked eyes. Everyday occurrences are utilized to advantage to describe the properties of light. The discussion shows the benefits derivable form from light waves.

Chapter 3, Electromagnetic waves, Wider Scope, focuses on the fundamentals of electromagnetic waves. Each form is being discussed broadly for better reference. The advantages that these forms can be used to enhance life, especially, in the field of medicine and engineering.

Chapter 4, Radiation and the Earth, gave importance to the transformation of other forms of energy to other forms of energy. Electromagnetic forces and electromagnetic induction are also tackled in this chapter. The discussion also shows the beneficial effects that can be derived from and the environmental hazards and perils that nuclear energy brings.

Chapter 5, Electrical waveforms and Signals, focuses on the use of electrical energy in the household. Concepts are tackled with the transformation of electrical energy to other forms, electrical energy consumption. The beneficial concepts that can be derived from electrical energy and the hazards that it may bring to every household that are consuming electrical energy.

Chapter 6, Earthquake Waves, centers on the discussion about earthquake. It gives information about the nature of earthquake waves. People must be aware of the intensities brought by earthquake to give awareness and warn the people about the destructions it may cause to both life and properties.

Chapter 7, Sound Waves, starts with the nature of sound and the energy it carries that brought different applications that may aid in the progress of the human life when it comes to technology.

Chapter 8, Applied Science, denotes the nature of applied science and be aware of the branches that Applied Science has. The benefits and hazards of technology are also discussed in this chapter as well as an understanding of technology invading the society. Technology and courses are also provided for the students to infer and may help in deciding what course they have to take.

Chapter 9, Energy Waves and Communication Technologies, tackles how communication evolved via sound waves. In this chapter, the history of the creation of telephone was discussed and how the telephone was considered as a major breakthrough in the field of science and technology. Concurrently to the creation of the telephone is the launching of electronic transport network. The telephone system or how the telephone transmits information from wire to wire.

Chapter 10, Wireless Communication Technology, discusses the launching of wireless communication through radio transport network. Wireless technological advances are also presented to keep the society abreast with advances in science and technology. How communication keeps the people in touch who are separated by geographical distances.

A historical concept of every topic in every chapter was given emphasis to provide the development of physical theories and technological advances is included to highpoint the humanistic aspect of Science.

The author hopes that this book will be an instrument for every student to become more conscious of the significance and relevance of science, specifically physics and technology in their everyday life.

Editor

CHAPTER 1

WAVE MOTION

INTRODUCTION

In the physical world, waves occur in all aspects. When you beat a drum with stick, sound waves carry the noise around you. Sound waves can be detected by the ears. When you stay outside and exposed to the sun for too long, your skin can be burned by ultraviolet waves. When tectonic plates move, earthquake waves carry energy that can be destructive to mankind depending on the intensity it brings. Turning the lamp on causes light waves to occupy the room. Some waves are visible to the naked eye; some are heard and can be felt. Though waves differ in many ways, they still all share the properties of wave motion.

Things in the environment shake, they fluctuates and tremble. Even atoms wiggle back and forth. These vibrations of objects are very useful to the mankind. Light in the form of waves enable us to see the beautiful world. Sound waves make us hear the sounds produced by all the things and people around us. When we talk is also an example of waves, the vibration produced in the voice box produces sounds. Wave is an intervallic vibration that the direction of the movement is away from the cause of shaking and transmits energy with it. Wave motion moves from a point of reference to another through a medium which displaces the particle of the vibrating body. Waves are composed of fluctuations or disturbances of such quantities in almost constant locations. A wave produced by slinky is a sample of longitudinal wave, when it is strained from both ends and seized at rest, it assumes a natural position known as balance or its equilibrium state or repose point. Once a force applied on it and caused the other end to shake back and forth, it can be observed that once it moved, it returned to its original position or in balance state or repose position. When the coils of the slinky displaced to a specific position and then recurring it to its original position, then disturbances are present in the coil spring. It can be observed then that disturbance moved over the coils starting on one end to another end. A slinky pulse is created through the movement of the coils from a given point to other point. Where in intervallic vibrations can be observed when the coils moved back and forth. These pulses are perceived to be the disturbance travelling over a medium from a specific origin to its destination. . A medium is another term commonly heard when wave is discussed. What is this medium and how important it is in propagation of the waves? A medium is a material or particle that conveys waves. Medium is not a wave nor creates waves; it is just a line that waves pass through.

It is very important in order for the waves to propagate because there are categories of waves that require medium in order to exist. Medium consists of particle that is interacting with each other.

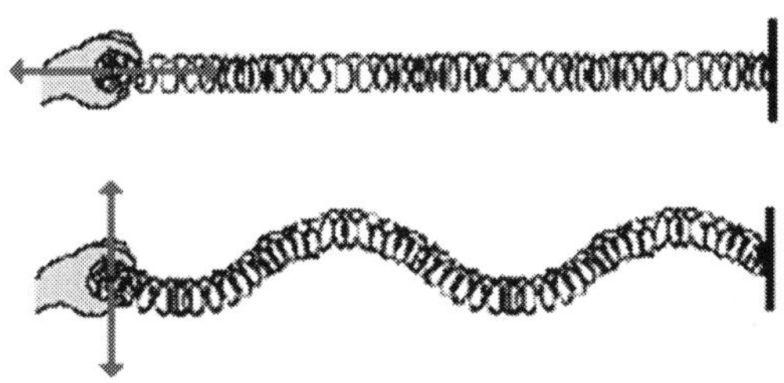

Slimkywaves can be made bt vibrating the first ciol back and forth in either a horizontal or a vertical direction.

Figure 1.1: shows that oscillating coil springs can cause waves which travels through the coils from one place to another repetitively.

Waving of hand is also an example of wave. Waving is a common gesture that people do to catch someone's attention or suggests a farewell. The repetitive motion of the hand while waiving is called a vibration. A vibration causes wave motion. Whenever you observe a wave, vibration would always be its source. However, vibration cannot be categorized as a wave for studies showed blurred line for the phenomenon is lacking of essential and ample features to define it as a wave.

Wave equation is one way to describe a wave; it explains how the trembling of the particles proceeds over time. The type of wave determines which mathematical equation must be used.

MATHEMATICAL EQUATION OF ONE-DIMENSIONAL WAVES

Let us say that a pulse of traveling waves in a string has a single dimensional spatial. Assume that a transverse wave is traveling. The equations below are considered useful in solving one dimensional waves that are travelling.

Calculation given in a wave with one-dimension

$$\frac{\partial^2 \psi}{\partial x^2} = \frac{1}{v^2}\frac{\partial^2 \psi}{\partial t^2}$$

To determine the identification of a wave, boundary or limited conditions are given to the equation

$$\psi(0,t) = 0$$

$$\psi(L,t) = 0$$

And initial conditions

$$\psi(x,0) = f(x)$$

$$\frac{\partial \psi}{\partial}(x,0) = g(x)$$

Using the solution formulated by d'Alembert can solve the one-dimensional wave equation, in addition to the usage of the transformational method proposed by Fourier, or through splitting of variables.

It was 1746 when d'Alembert formulated his solution, then in 1748, just after two years using the d'Alembert's solution, that method was further explained and extended by Euler.

Let

$$\xi \equiv x - vt$$

$$\eta \equiv x + vt.$$

From the series of rule,

$$\frac{\partial^2 \psi}{\partial x^2} = \frac{\partial^2 \psi}{\partial \xi^2} + 2\frac{\partial^2 \psi}{\partial \xi \partial \eta} + \frac{\partial^2 \psi}{\partial \eta^2}$$

$$\frac{1}{v^2}\frac{\partial^2 \psi}{\partial t^2} = \frac{\partial^2 \psi}{\partial \xi^2} - 2\frac{\partial^2 \psi}{\partial \xi \partial \eta} + \frac{\partial^2 \psi}{\partial \eta^2}$$

Then it transforms to

$$\frac{\partial^2 \psi}{\partial \xi \partial \eta} = 0.$$

This is the form of equation to be used in any solution

$$\psi(\xi,\eta) = f(\eta) + g(\xi) - f(x+vt) + g(x-vt),$$

Where f and g are *any* functions. These functions signify two forms of waves that are transporting in a direction against with each other, f in the negative x direction and g in the positive x direction.

Through applying Fourier transform on each sideways the wave equation for one-dimensional can be can also be answered,

$$\int_{-\infty}^{\infty} \frac{\partial^2 \psi(x,t)}{\partial x^2} e^{-2\pi ikx} dx = \frac{1}{v^2} \int_{-\infty}^{\infty} \frac{\partial^2 \psi(x,t)}{\partial t^2} e^{-2\pi ikx} dk,$$

Fourier transform derivative identity helped in identifying the given by

$$(2\pi ik)^2 \psi(k,t) = \frac{1}{v^2}\frac{\partial^2 \psi(k,t)}{\partial t^2}$$

Where

$$\psi(k,t) \equiv F_x\left[\psi(x,t)\right](k)$$

$$= \int_{-\infty}^{\infty} \psi(x,t) e^{-2\pi ikx} dx.$$

The calculation involved is

$$\psi(k,t) = A(k)e^{2\pi ikvt} + B(k)e^{-2\pi ikvt}$$

Applying the opposite Fourier transform offers

$$\psi(x,t) \equiv \int_{-\infty}^{\infty} \psi(k,t) e^{2\pi ikx} dk.$$

$$= \int_{-\infty}^{\infty} \left[A(k)e^{2\pi ikvt} + B(k)e^{-2\pi ikvt}\right] e^{-2\pi ikx} dx$$

$$= \int_{-\infty}^{\infty} A(k)e^{-2\pi ik(x-vt)} dk + \int_{-\infty}^{\infty} B(k)e^{-2\pi ik(x+vt)} dk$$

$$= f_1(x-vt) + f_2(x+vt),$$

Where

$$f_1(u) \equiv F_k\left[A(k)\right](u) = \int_{-\infty}^{\infty} A(k)e^{-2\pi iku} dk$$

$$f_2(u) \equiv F_k\left[B(k)\right](u) = \int_{-\infty}^{\infty} B(k)e^{-2\pi iku} dk.$$

The calculation is still under to all other original and boundary conditions.

Splitting the variables via trial solution can answer the wave equation of one-dimensional wave pattern.

$$\psi(x,t) = X(x)T(t).$$

The given will be

$$T\frac{d^2X}{dx^2}=\frac{1}{v^2}X\frac{d^2T}{dt^2}$$

$$\frac{1}{X}\frac{d^2}{dx^2}=\frac{1}{v^2}\frac{1}{T}\frac{d^2T}{dt^2}=-k^2$$

Therefore the equation for X is

$$X(x)=C\cos(kx)+D\sin(k,x).$$

Modifying above equation provides

$$\frac{1}{T}\frac{d^2T}{dt^2}=-v^2k^2\equiv-\omega^2,$$

Therefore the calculation for T is

$$T(t)=E\cos(\omega t)+F\sin(\omega t),$$

Where $v\equiv\omega/k.$. Considering the boundary conditions $\psi(0,t)=\psi(L,t)=0$ to (◇) provides

$$C=0\,kL=m\pi,$$

Where m is an integer, Persisting (◇), (◇) and (29) back in for ψ in (◇) stretches, for a specific value of m,

$$\psi_m(x,t)=\left[E_m\sin(\omega_m t)+F_m\cos(\omega_m t)\right]D_m\sin\left(\frac{m\pi x}{L}\right)$$
$$\equiv\left[A_m\cos(\omega_m t)+B_m\sin(\omega_m t)\right]\sin\left(\frac{m\pi x}{L}\right)$$

Its original condition $\psi(x,0)=0$ then gives $B_m=0$, so (31) turns out to be

$$\psi_m(x,t)=A_m\cos(\omega_m t)\sin\left(\frac{m\pi x}{L}\right)$$

The overall explanation is a sum over all possible values of m, so

$$\psi(x,t) = \sum_{m=1}^{\infty} A_m \cos(\omega_m t) \sin\left(\frac{m\pi x}{L}\right)$$

By using linear transformation of sines again,

$$\int_0^L \sin\left(\frac{l\pi x}{L}\right) \sin\left(\frac{m\pi x}{L}\right) dx = \frac{1}{2} L\delta_{lm},$$

Where δ_{lm}, Kronecker delta can be explained by

$$\delta_{mn} \equiv \begin{cases} 1 & m = n \\ 0 & m \neq n \end{cases},$$

Gives

$$\int_0^L \psi(x,0) \sin\left(\frac{m\pi x}{L}\right) dx = \sum_{l=1}^{\infty} A_l \sin\left(\frac{l\pi x}{L}\right) \sin\left(\frac{m\pi x}{L}\right)$$

$$= \sum_{l=1}^{\infty} A_l \frac{1}{2} L\delta_{lm}$$

$$== \frac{1}{2} LA_m,$$

So we have

$$A_m = \frac{2}{L} \int_0^L \psi(x,0) \sin\left(\frac{m\pi x}{L}\right) dx.$$

Calculation of A_ms for precise original spins is resulting in the Fourier sine series section. It is been found out thatB_m=0, therefore the equation for the string's motion (◇), with

$$\omega_m \equiv vk_m = \frac{vm\pi}{L}$$

Is

$$\psi(x,t)=\sum_{m=1}^{\infty}A_m\cos\left(\frac{vm\pi t}{L}\right)\sin\left(\frac{m\pi x}{L}\right)$$

Where the A_m constants are given by (◇).

One-dimensional wave that is restrained

$$\frac{\partial^2\psi}{\partial x^2}=\frac{1}{v^2}\frac{\partial^2\psi}{\partial t^2}+b\frac{\partial\psi}{\partial t},$$

Given boundary conditions

$$\psi(0,t)=0$$
$$\psi(L,t)=0$$

Initial conditions

$$\psi(x,0)=f(x)$$
$$\frac{\partial\psi}{\partial t}(x,0)=g(x)$$

And the added limit

$$0<b<\frac{2\pi}{Lv}$$

It can be answered as a Fourier series.

$$\psi(x,t)=\sum_{n=1}^{\infty}\sin\left(\frac{n\pi x}{L}\right)e^{-v^2bt/2}\left[a_n\sin(\mu_n t)+b_n\cos(\mu_n t)\right]$$

Where

$$\mu_n \equiv \frac{\sqrt{4v^2n^2\pi^2 - b^2L^2v^4}}{2L} = \frac{v\sqrt{4n^2\pi^2 - b^2L^2v^2}}{2L}$$

$$b_n = \frac{2}{L}\int_0^L \sin\left(\frac{n\pi x}{L}\right) f(x)\,dx$$

$$a_n = \frac{2}{L\mu_n}\left\{\int_0^L \sin\left(\frac{n\pi x}{L}\right)\left[g(x) + \frac{v^2 b}{2} f(x)\right] dx\right\}.$$

WAVEFORMS

Wave forms deal with shapes and forms of any disturbance that is traveling from a specific medium. Sometimes the shape and the form of a wave do not observable or portrayed directly on the medium where wave representation is present. Oscilloscope is an instrument used to determine the shape or the signal that the disturbance traveling on a medium. The change in electrical signal observed by the oscilloscope such as voltage and shapes of the waves are continuously recorded in a graph in a calibrated scale. Then the waveforms that are observed through oscilloscope are investigated for their inhibited properties such as amplitude, frequency, period, interval of time and others. Since waveforms are recorded in a graph, its amplitude is represented and measured on the vertical position which is the y axis and time will be on horizontal side which is the x axis.

- Sine wave. The sine wave is observed to be following a waveform that is a trigonometric sine function in respect with time. It is also known as the sinusoidal waveform.

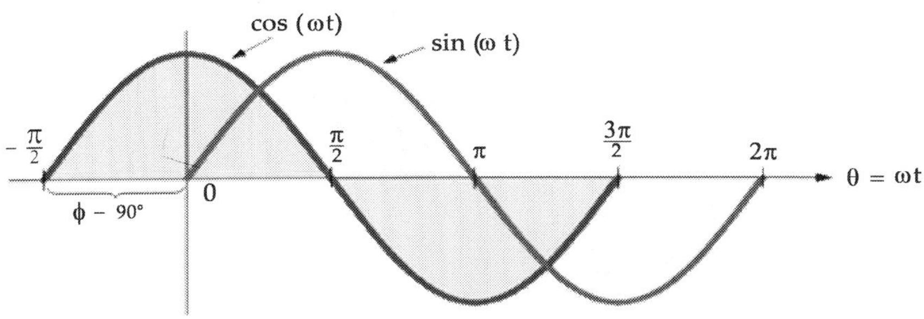

Figure 1.2: Sine waveform.

- Square wave. This kind of waveform is used to embody digital information. It contains odd divergent infinite series that reduced in -6 dB/octave.
- Triangle wave. It represents odd divergent infinite series that reduced at -12 dB/octave.
- Saw tooth wave. Waveform that looks like a row of teeth of a saw. It is used as bases of display scanning and original point for sound synthesis. Saw tooth waveform displays a constant period that has odd and even divergent infinite series or harmonics that reduced at -6 dB/octave.

The disintegration of periodic waveforms, as described by Fourier series, which any periodic waveform can be a result of the sum of a finite set of fundamental and divergent infinite series components or harmonics. Other examples of waveforms are called composite waveforms which can be defined as the combination of two or more sinusoidal waveforms.

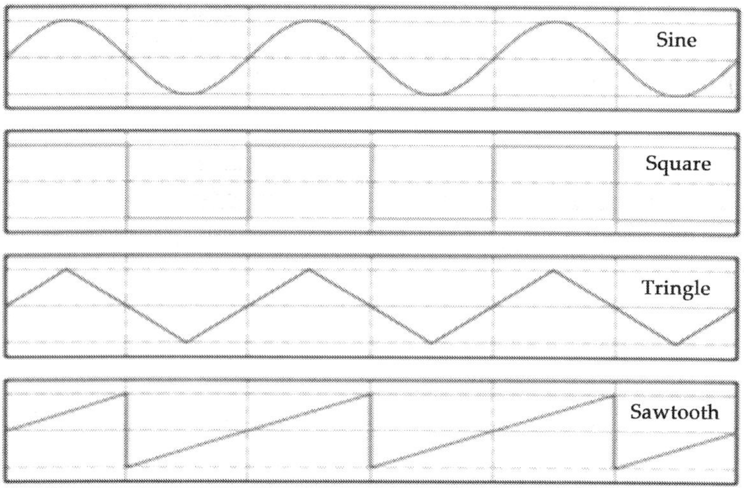

Figure 1.3: Sine, square, triangle and saw tooth waveforms.

HOW ENERGY IS TRANSFERRED THROUGH WAVES?

Waves are created from the vibration of the medium and the energy from the source causes the vibration. As the medium vibrates the energy is transferred from a point of reference to another place. A constant force is present which act on the particles of the medium that restores them from their origin. Let us take water waves as an example. Once the water was touched billows are formed. The cause of the energy is the hand that taps the water. Rap the surface of the water softly creates several billows but tap it with greater energy then the billows increased in numbers as well. To test whether the waves carry energy, a paper boat is placed on the water, then tap the water softly. As a result the paper boat was transferred from the reference point to another point. However tapping the water with greater energy, the paper boat is displaced from its origin to another place faster. Therefore, the greater energy from the source the greater energy the waves carry.

Waves are observed to move through bodies or water such as seas or lake; however, the water keeps on yielding to its rest position. The particles of water are not transferred; what is being transferred through the medium is the energy alone. The evidence of this phenomenon is the presence of the water in the center of the sea. The water particles did not transported to the shore from the midst of the ocean, what is being transported to the shore is the energy carried by the waves. As observing a bird that rests on the water, it moves up and down as waves pass by. Even displaced into another position, it will still return into its original position. The bird keeps on returning to its initial position. It is evident that the water which the bird has rested on did not transported to shore because the bird remains on its original position. It is observed that only the energy that passes through the water wave has been transferred to the shore.

STRUCTURE OF A WAVE

The pulse of a periodic wave follows an even series of waves. Waveforms with specific shape are recurring in regular intermissions. Position A which is called crest displayed the uppermost point of a wave. It is the point in the mediating object that extended to the maximum amount of affirmative or upward transposition from the initial location. The trough, position B, is the lowest point or the point that shows the highest amount of negative or downward transposition from the initial location. Wavelength, from position A to B, refers to the space between two consecutive matching parts of the wave. Wavelengths can be seen as two successive crests or troughs. Amplitude is the extreme transposition of trembling particles on both sides of regular position of the passing waves. It refers to the wave height from a trough to crest. Frequency refers to the number of waves that are passing in a medium over time. Frequency is related to wavelength in terms of the higher the frequency the shortest the wavelength of a certain wave. Period is the quantity of time which a wave such as crest and trough used to travel from the initial point to final point.

MECHANICAL WAVES VS. ELECTROMAGNETIC WAVES

There are different shapes and forms which waves inhibit. They all share the same distinguished characteristics, some are observable and some are not. The ability and inability to transfer energy through a vacuum is another way to categorize a wave. Mechanical waves and electromagnetic fall under this category.

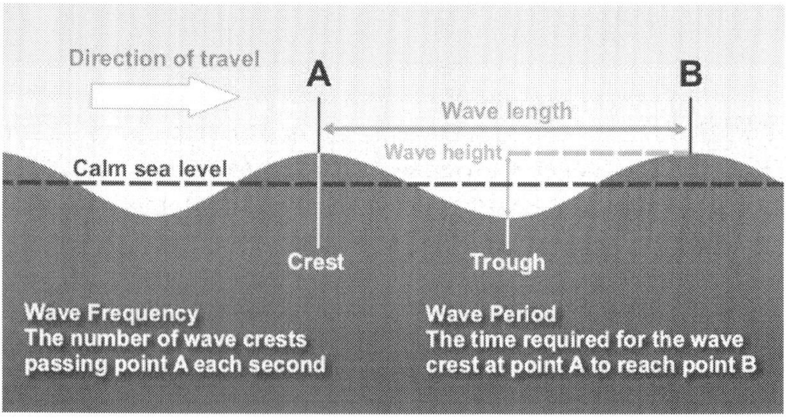

Figure 1.4: Anatomy of a wave.

Mechanical Waves

Mechanical waves need a medium in order to transfer energy from initial point to another. A medium is required in all waves fall under mechanical waves; it is required to propagate waves. Mechanical waves have the inability to transmit energy via empty space or vacuum. One of the best examples of mechanical waves is the sound wave. Sound waves cannot pass through an empty space because there is no sound in outer space. Waves produced by a coiled string, surface waves (water), and earthquake waves, and propagated waves through a rope are other examples of mechanical waves; these examples need a medium for their wave propagation. A slinky wave involves the coiled spring in order to be produced; a surface wave requires water; earthquake waves require land to propagate.

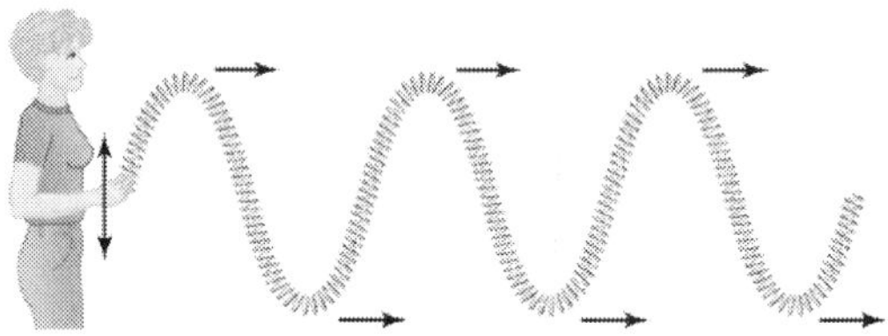

Figure 1.5: Waves propagated through a jump rope.

Figure 1.6: Surface waves require water in order to propagate.

Transverse Waves

Transverse waves are waves produced from the vibration that is perpendicular to the direction of the movement of the medium. The waves produced in shaking the rope up and down are called transverse waves because they display vibration on the separated parts of the rope perpendicular to the path in which the waves are traveling. Doing the activity where you will flatten the rope and put it above a long flat surface, then hold one end of the rope and vibrate it up and down. A series of pulse are observed; waves that move toward the direction of the energy exerted from the source of the wave. The energy exerted by the hand is the basis of the wave and travels to the medium, in this example; rope is the medium of the wave propagation. The movement of the hand in creating the pulse is the same with the motion of the waves with respect to the source of the energy.

Transverse waves are always categorized by the movement of its particles that are perpendicular to wave motion. Transverse waves fall under the category of mechanical waves since a medium is needed in order for these waves to be created.

Longitudinal Waves

In case of longitudinal waves, waves travel back and forth with respect to the movement of the particles of the medium which move parallel to the source of the energy. The constituent part of the medium travelled in the same direction or parallel to the direction of energy transport. Slinky wave is an example of longitudinal waves, when it is stretched back and forth from one end while the other end is attached to a wall. Longitudinal waves in slinky moves in horizontal direction if the coiled spring is stretched forward and backward repeatedly. The coiled spring with respects to its coils move horizontally therefore it has a part where coils are compressed and some parts that are decompressed.

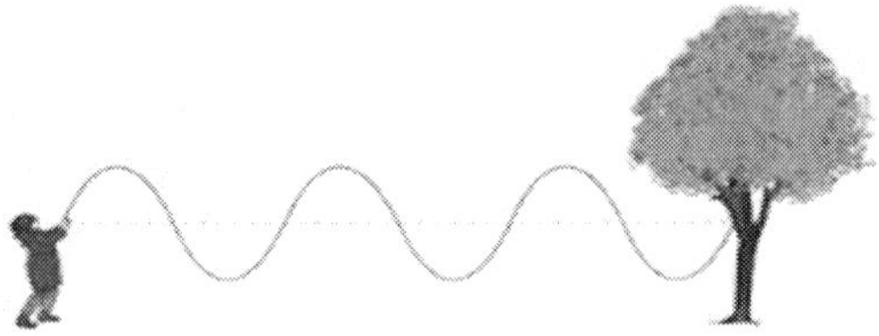

Figure 1.7: Transverse waves created through a rope knotted in a fixed object.

A longitudinal wave that is traveling in a medium that display maximum density is called compression. Rarefaction or the decompressed region of the particles of the medium maximizes the length of the space between the coils. A point on a medium with minimum density which a longitudinal wave travel is known as rarefaction. The compressed region is represented by the points B and D while rarefaction or the decompressed area is represented by points A and C. It can be observed that alternating pattern of crests and troughs that are perpendicular to the wave motion define transverse waves, while a longitudinal wave is composed of alternating pattern of compressions and rarefactions that move parallel to the energy exerted from the source.

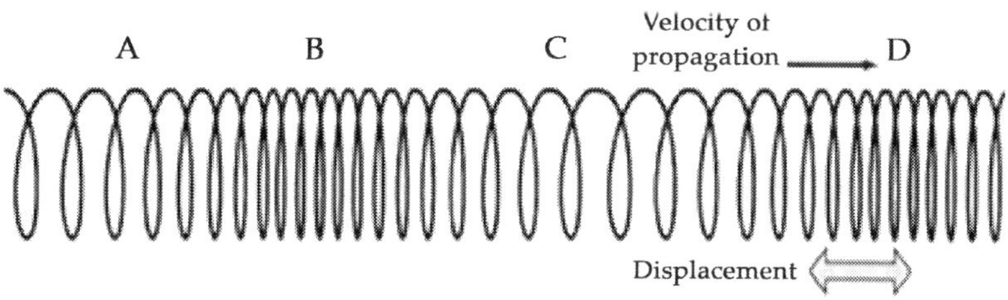

Figure 1.8: Longitudinal waves displayed in a slinky.

Surface Waves

A wave in which particles follow a circular pattern is known as a surface wave. Surface waves can be both transverse and longitudinal waves. The motion of water molecules display mixed transverse and longitudinal displacements when the waves move, the result is that the molecules at the surface move in almost circular paths. The molecules of the water are moved in horizontal and vertical path from its reference point. When you hit the surface of the water in a basin, waves are formed and subsequent crests of the water are observed.

There are two main types of waves these are body waves and surface waves. Waves that can travel through the earth's crust is called as body waves, however the waves that travel on the surface of earth's like ripples on water is known as surface waves. Earthquakes can be classified both body waves and surface waves. It discharged seismic energy. When a surface wave travels in earth's crust it is perceived to have lower frequency than body waves and as an outcome it is detected by the seismogram. They are almost entirely accountable for the destruction and devastation linked with earthquakes. Nonetheless, as the earthquake becomes deeper the damage and the strength brought out by surface waves are reduced. Surface waves are classified into two, namely Love waves and Rayleigh waves. In 1911, a British mathematician, A.E.H. Love toiled out the mathematical model for this type of waves. Love wave was named after A.E.H Love. It moves on the ground side to side and is known to be the fastest surface waves. John William Strutt, Lord Rayleigh who foreseen the existence of this type of wave through mathematical equation. It was on 1885 when the second type surface wave was discovered, named after Lord Rayleigh; the Rayleigh wave. The movement of Rayleigh waves is almost the same with how waves roll across the bodies of water. Due to its movement the ground moves up and down, and side-to-side in the same direction that the wave is moving. Rayleigh wave causes most shaking that can be felt from an earthquake.

Sound Waves

Sound wave is another type of wave sunder mechanical waves. It is a pattern of disruption that travels via medium such as air, water or any liquid and solid matter. Sound waves travel in solid objects best and least energy was transmitted through gas state matter because its particles move rapidly while in solid objects, the particles are tightly compacted. Sound waves cannot be transmitted through a vacuum; there is no sound in outer space. It is also a kind of pressure wave where in it has two variations, the region of increased pressure on a sound wave is called a compression (or condensation), and the rarefaction (or a dilation) a region of decreased pressure on a sound wave.

Transverse waves, longitudinal waves, surface waves and sound waves are all categorized as Mechanical waves for they cannot travel in a vacuum. All of them require medium in order to proliferate. A medium, which can be solid, liquid or gas, is an object where waves pass through.

Electromagnetic Waves

Electromagnetic waves required no medium in order to travel or propagate. It can travel through an empty space or vacuum. These waves are outcomes of vibrating charged particles. The ultraviolet waves that the Sun produced are electromagnetic waves that pass through an empty space to the Earth. It can be said that without the capacity of these electromagnetic waves to transmit through a vacuum, life would be impossible on earth. An electromagnetic wave conveyed its energy through an empty space or in a vacuum at a speed of 3.00×10^8 m/s (a speed value commonly represented by the symbol c). The dissemination of an electromagnetic wave through a material medium take place at a net speed which is less than 3.00×10^8 m/s.

The transfer of energy of electromagnetic waves includes absorption and reemission of the wave energy by the atoms of the material. When the energy of electromagnetic waves intrudes the atom, it absorbs the energy.

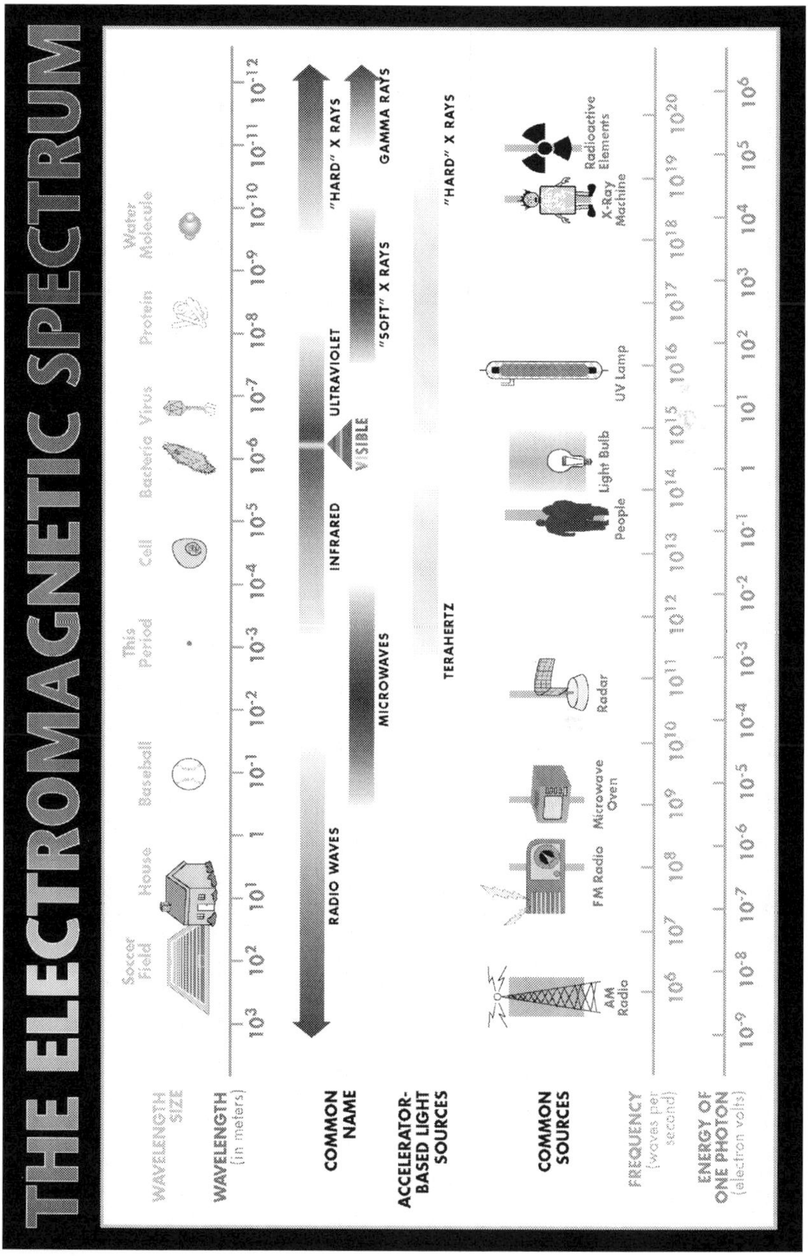

Figure 1.9: The Electromagnetic Spectrum.

Different types of electromagnetic waves with their range of frequencies and wavelength are shown in the Electromagnetic Spectrum. The range of speed of all electromagnetic waves are approximately equal to 300, 000, 000 m/s; light waves have similar range of speed. It is also shown in the figure that the greater the frequency of the waves the shorter its wavelength. The longer the wavelength the lower the frequency of the waves.

Types of Electromagnetic Waves

1. Radio waves have the lowermost frequency among the types of electromagnetic waves; it also suggests that they have the least amount of energy. The wavelengths of radio waves measured around 11 inches, just like a size of football. When a person tuned in to a radio station, he or she is heeding to a definite frequency of radio waves. The frequencies of the radio station are represented by numbers, just like in AM stations they are represented by numbers ranging from 520 to 1610, at thousands of hertz per second, or kHz (kilohertz). The range of radio frequencies with the highest range of 107.9 and down to 87.0 million hertz per second, called megahertz or MHz represented the FM stations.
2. Microwaves. The waves of this type of electromagnetic wave measured less than 11.8 inches long. Microwaves are also used to transmit signals to radar, television and satellites; they are not just for heating left over food for consumption. Microwaves display higher frequencies consisting of cycles that are taking place in every second. It is composed of billions or trillions of cycles.
3. Infrared waves have higher frequency compared to microwaves. The most useful things about infrared are; they are used to rheostat television remote controls and for thermal imaging, for example the night vision goggles which used infrared thermal imaging in order to see things at night or very dark room or places, second is that when sunlight gives energy through sun rays, the energy that the rays contained is an infrared radiation from the

sun. Because of their high frequencies, the wavelengths of infrared are so minute; they are only hundredths or thousandths of an inch,

4. Visible light is the only electromagnetic wave that can be seen through the naked eyes. The different colors of the rainbow can be detected by the eyes, especially when it passes through a prism. Violet, which wavelengths measure around 400 nanometers, has the shortest wavelengths thus it has greater energy and red, which wavelengths measure around 700 nanometers, having the longest wavelengths considered to have a lowest amount of energy.

5. Ultraviolet (UV) light has shorter wavelengths than visible light. UV waves are visible to some insects, such as bumblebees; however human eyes failed to detect UV rays though it is felt by the skin when someone stays too long under the sun.

   The Sun is a source of the full spectrum of ultraviolet radiation, which is commonly subdivided into UV-A, UV-B, and UV-C. These are the classifications most often used in Earth sciences. UV-C rays are the most harmful and are almost completely absorbed by our atmosphere. UV-B rays are the harmful rays that cause sunburn. Exposure to UV-B rays increases the risk of DNA and other cellular damage in living organisms. Fortunately, about 95 percent UV-B rays are absorbed by ozone in the Earth's atmosphere.

6. X-rays have a high frequency waves, and carry a lot of energy. They will pass through most substances, and this makes them useful in medicine and industry to see inside things. It used by the doctors to treat illnesses without doing the surgery. X-Rays are also used in airport security checks, for security purposes, to see inside your luggage. They are also used by astronomers - many objects in the universe emit X-rays, which we can detect using suitable radio telescopes. Lower energy X-Rays don't pass through tissues as easily, and can be used to scan

soft areas such as the brain.

7. Gamma rays consist of photons in the highest observed range of photon energy. Denoted by a lower-case Greek letter gamma (y or G) a powerful electromagnetic radiation that came from radioactive decay of atomic nuclei. Gamma radiation was discovered by Paul Villard, a French chemist and physicist, in 1900 while studying radiation emitted by radium. Ernest Rutherford named this radiation gamma rays later on 1903. Rutherford had previously discovered two other types of radioactive decay, which he named alpha and beta rays. Gamma rays are hazardous because they can ionize other atoms. Gamma radiation produces when an atomic nucleus decays from a high energy state to a lower energy state. Gamma rays typically have frequencies above 10 exahertz (or >1019 Hz), and therefore have energies above 100 keV and wavelengths less than 10 picometers (10−11 m), which is less than the diameter of an atom.

Sound waves, water waves, earthquake waves, and waves in stretched string propagate through a medium are called Mechanical waves, while electromagnetic waves such as radio waves, microwaves, visible light, infrared, ultra violet waves, x rays and gamma rays, do not require a material medium for their passage.

THE SPEED OF A WAVE

In order to get the speed of a moving body, the distance that the moving body traveled from its origin to the destination must be determined and how fast the body reached its destination. It is usually expressed as the distance traveled over time of travel. In waves, to determine its speed, identify the distance traveled by specific affirmative or with maximum displacement point on the wave (such as a crest) divided by set of time interval.

For instance, when a crest of a rope wave moves a distance of 10 meters in 2 seconds, the calculated speed of the waves is 5m/s. On the other hand, if the crest of a rope wave moves at a distance of 20 meters 2 seconds (the same amount of time), then the speed of the rope wave is

10 m/s. The speed calculated with greater distance with the same period of time is perceived to be the faster waves.

For reflection phenomenon, where in a wave come across at the end of a medium and the incidence of another medium. For example, waves are generated by two persons holding each end of the rope. As the first person vibrates the end of the rope then waves travel through the rope to the other end. Reflection phenomena happened when the second person holding the other end send the vibration back to the initial position. This time, a wave that experiences reflection, simply goes back to its initial position without changing the manner how it passed through the medium. It is observed in the waves created through a rope, the occurrence of vibration can be portrayed that it is moving back from its origin. In the case of a slinky wave, the disturbance can be seen traveling back to the original end; therefore it has doubled its distance.

Sound waves commonly produce reflection phenomena. When you yell near a mountain, echo of your voice is being transported back. The sound wave then travels through the air and bounced back off the mountain wall and yields it to its original position (you). As an outcome you hear the echo of your voice and that is an example of reflected sound waves.

For example, Rico is walking 100 m away from an alley and calls his friends who are walking ahead of him. Once he yells out the name of his friend an echo of his own voice was heard 2 seconds later. Can you calculate the speed of the wave?

The speed of the wave is 100 m/s. In reflection, the distance travel by the wave will be doubled. The distance travel by the sound to the valley is 100 m in 2 seconds. One hundred meters divided by 2 seconds is 50 meters per second then another 50 m/s when the wave travels back from the valley from to Rico.

HOW WAVES MOVE

Boundary behavior is the behavior of the wave when it reached the limit of a medium. Boundary is the edge of the two media, thus when one media ends, another media begins. Just like in the example given earlier when a person let out a yell to the canyon, the waves move back to the person. Same through with the slinky being held by the two persons. As the other slinky was moved by the person then reach its end, then the other person who holds the Slinky's other end will move it backward. Then the distance covered by the waves are doubled when it reached its end and when it goes back to the other person and vice versa.

Fixed End reflection

Fixed end reflection is best demonstrated by a rope tied to a rigid object. When the disturbance reached the other end of the rope, which is tied to the rigid object, the energy on the last particle of the rope will be unable to move the vibration back because it is attached to a fixed object. This end of the rope is referred to as a fixed end reflection is observed in the other end of the rope which wave cannot travel back from its initial position. A pulse called incident pulse is presented at the rope's left end and traveled through the rope towards the medium's right end; this how waves behave in fixed end reflection. Incident pulse is created because of the occurrence of the boundary with the rigid object. Two things happened when incident pulse reached its boundary.

A. Reflected pulse is then created when a reflected part of the energy being carried by the pulse returns to the rope's left end. It is a wave or vibrations that transmitted back to the left end of the rope after rebounding back from the rigid object that the rope was tied into.

B. Vibration of the pole as the pulse transported the part of the energy it carried to the rigid object where the rope is tied into.

Free end Reflection

Free end reflection can be shown when a cord is affixed to a rigid object but this time instead of tied securely, ring will be used. This ring will make the rope moves loosely that is why when the remaining particle of the medium reached the end of the rope it will be able to move the object because it is attached to the object freely. And it refers to free end.

REFLECTION, REFRACTION, AND DIFFRACTION

Reflection

Light is the reason why all the things in this world can be seen. Light moves as a wave then reflect and bouncing off light for the objects to be seen. When the light bounced off a material, it refers to reflection. A wave's reflection is defined as the process of the wave hitting a surface and by which a wave hits an object and spring back at different location. However, the reflection of light on a mirror is different in a plain wall. When a surface was hit by a light waves, opaque reflection is created; light then will be reflected from that surface mostly. Incident ray is the wave moved toward the surface and then reflected ray is generated when it bounces off that surface. Law of reflection stated that the incident angle is congruent to that reflected angle in a way that waves usually approach the barriers with the angle equals to the angle at which waves bounce off the barrier. It is also stated that if it let you to draw a line between the two rays perfectly then the incident angle is congruent to the reflected angle.

Types of Reflection

1. Specular reflection is a reflection of all light rays from a surface of a mirror bounce off having the same angle. The law of reflection is true, when you use mirrored surface to regular surface it is the surface that changed and not the law.

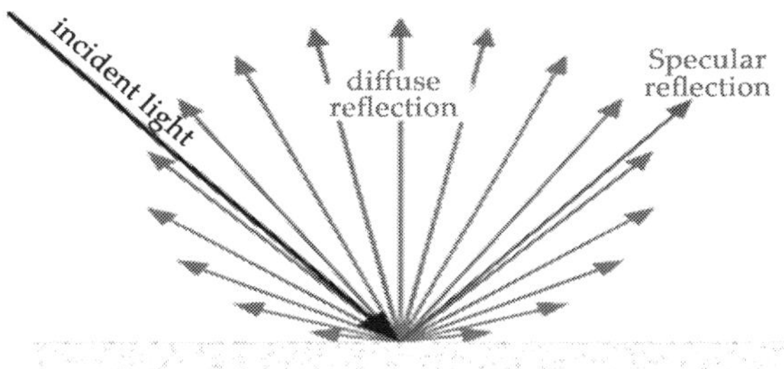

Figure 1.10: Types of reflection.

2. Diffuse reflection defined as when waves such as light waves smash on several deficiencies and bounce off at different reflected surface. I have something to do with a smooth surface but if you take a closer look there are millions of tiny imperfections. However, because of these tiny imperfections, a light wave doesn't hit the surface that we see.

Refraction

A wave passing from one medium to another that involves change in the direction of waves is known as refraction. The bending of the path of the waves complemented by a change in speed and wavelength of the waves simply defined refraction. It can be observed when a pencil is placed in a glass half-filled with water and as you take a look at the pencil it is bent, and it seems to be traveling more vertical to the surface. When waves are traveling from different depth of bodies of water from a shallow water to deep water, then waves appeared bent against other direction.

Figure 1.11: An image showing how refraction of light happens

Diffraction

When waves pass through a hole or around a blockade in their path there will be a change in direction and that is what we called diffraction.

Observing the path of the water waves as they encounter barriers is best demonstrated diffraction. The waves are seen to pass around the barrier into the regions behind it; subsequently the water behind the barrier is disturbed even though the waves seem to pass around the obstacles. An increasing wavelength increases the amount diffraction and decreasing wavelength decreases the amount of diffraction. So when the waves' wavelengths are larger than the obstacle, there will be evident that diffraction may occur.

Water waves are observed to be diffracted in a dock, the waves bend around small barriers such as small boats and they disturbed the water behind them. However larger barriers such as large boats unable the water waves to diffract because their wavelength is smaller than the boat. Diffraction of sound waves is greater when the sound waves come from a larger opening such as open doors, but sound diffraction around small holes in the wall is lesser due to the size of the holes that may hindrance the sound to diffract, as a result you cannot hear who are speaking or what do they talk about. Light waves also diffract when its waves fluke into obstacles or barriers with tremendously tiny wavelengths (like molecules that are up in the air hanging).

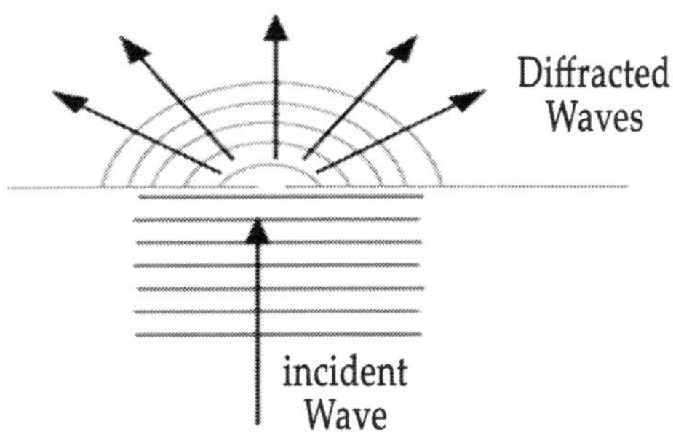

Figure 1.12: An image showing how diffracted waves behave.

DOPPLER EFFECT

Doppler Effect was proposed by Christian Doppler, an Austrian physicist, in Prague year 1842. Doppler Effect is defined by how sound intensity is changed when an ambulance is approaching, on its instant passing and its recession from an observer. A siren of an ambulance has lower intensity as it approaches compare to the actual passing of the ambulance. However the intensity of the sound of the siren is higher when it passes than when it recedes from the observer. The frequency received by the observer is higher during the approach compared to the emitted frequency, then the frequency received and transmitted are identical as the ambulance passes by and lower received and transmitted during the ambulance recedes.

The diagram below shows the Doppler Effect on Sound.

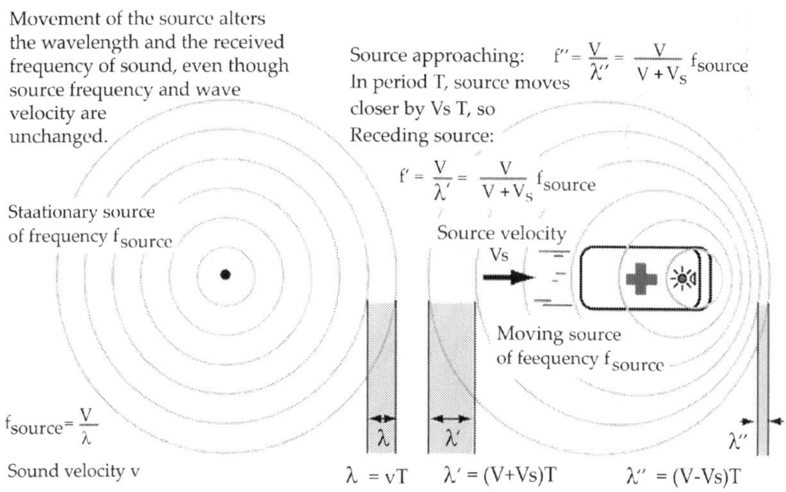

Figure 1.13: Doppler effect diagram.

Doppler Effect is when frequency of a wave-like signal - such as sound or light - depends on the movement of the sender and of the receiver some of its manifestations, we know from everyday life, such as a fire engine's siren abruptly changing pitch as the engine passes by; others are of interest in astronomy and astrophysics. The purpose of this spotlight text is to take a closer look at what the Doppler Effect is all about.

THE DIFFERENCE BETWEEN TRAVELING WAVES AND STANDING WAVES

Traveling wave is a type of wave pattern that is seen moving via a specific medium. When the wave is perceived roving through a medium, a crest moves perpendicularly among the particles of the medium. This crest is followed by a series of troughs and crests. Actually, sine wave pattern will continues moving in perpetual method till it bump into another wave along the medium or until it encounters a boundary with another different medium. Ocean wave is the most common traveling wave; traveling waves can be observed easier when it is not confine in small region or place. Traveling waves are also observed in a rope tied in a rigid object. If someone continued to shake the free end point of the rope it will generate traveling waves continuously. Traveling waves can be perceived as series of pulse as well. Meanwhile, standing waves follow a wave pattern that is perceived to be perpendicularly still. If traveling waves are not confined in region standing waves are limited in a given space in the medium and readily seen regular wave pattern produced within medium's motion. With right amount of frequency when the rope is vibrated from one end, pattern of waves is propagated. The intervention of the incident wave and the reflected wave may occur if there is a right amount of frequency; it pertains to a manner that there are precise points along the medium that appear to be standing still. Standing wave is a vibration of a system in which some particular points remain fixed while others between them vibrate with the maximum amplitude.

CHAPTER 2

LIGHT WAVE

INTRODUCTION

Light is the only electromagnetic wave that can be seen by the naked eye. Similar to all electromagnetic waves, light has the ability to pass through a vacuum. Various experiments on interference and diffraction had demonstrated the wave nature of light. It is transverse in nature which explained that it can travel through a medium as well; being not disturbed by gravity, it is espoused that light transports via medium in slower speed. This occurrence is explained by Christian Huygens on his mathematical wave theory of light; that light was released in all tracks through a medium called luminiferous ether as series of waves. Another evidence is through the pulse theory proposed by Robert Hooke, he recommended that the disturbance or vibrations made by light may possibly perpendicular to the track of its proliferation.

Visible light is another term that refers to light, visible because it can be observed by the human eye and accountable for the sense of sight. It is perceived to have wavelengths that measure in a range of 400 to 700 nanometers. It unveils properties of being a wave and a particle at the same time when light is transported then absorbed by photons which described as little packets. Visible light is a wave of energy that is also emitted by the Sun just like infrared and ultraviolet rays. It can be transformed into chemical energy by the leaves of the plants in a process known as photosynthesis.

Visible light makes living on earth wonderful. Aside from sustaining the life on Earth through giving heat and light, people can see the beauty that the nature offers and suppress the perpetual darkness on earth.

SIGNIFICANT HISTORICAL CONCEPTS OF LIGHT

Theories from Ancient Greece

Empedocles from fifth century hypothesized that everything that exists is comprised of the four elements called fire, air, water and earth. According to him, Aphrodite, the goddess of beauty, created the human eyes out of these elements. The sense of sight became possible when she ignited fire in the eyes, this fire lights the yes in order to see things. However, there is a debate that if Empedocles is right about his hypothesis then a person could see things at nighttime as clearly as in daytime. Therefore, he proposed a theory that explain the interaction between the eyes and the rays, and then the interaction between the rays and source of light which is the Sun.

A Roman named Lucretius, who believed in concepts of early Greek philosophers who studied atoms, shared his insights about light and heat. It is written on his book entitled "On the Nature of the Universe", that light and heat of the sun are composed of minute atoms which when pushed they miss no time in traveling via air through the direction of the force exerted on the. However, Lucretius' sights though related to the particle theory are not accepted. Lucretius was from the 55 BC and it was on second century when Ptolemy wrote his book entitled "Optics" which tackles the refraction of light.

It was then in 300 BC that another mathematician named Euclid studied the light's properties. He wrote "Optica" which explained that light transported in straight path and explained the law of reflection which he applied mathematics in studying them.

Descartes' Theory of Light

René Descartes (1596-1650) believed that light was a luminous body's mechanical property; he rejected the ideas of other scientists. It was on 1637 when he published his refraction of light's theory, which is presumed to be incorrect when he perceived that light transported in a denser medium is faster compared to the light transported in a less dense medium. His analogy with the behavior of sound waves made him arrived to this conclusion. However, although his concepts of relative speed was improper, he was correct in proposing that light moved like a wave and his conclusion that the speed of light in different media could explain the refraction of light.

Particle Theory

An atomist named Pierre Gassendi (1592-1655) suggested an atomic theory of light; it was only published after his death in 1660s.Then young Isaac Newton studied the work of Gassendi, conversely, chose Descartes' plenum's theory as basis of his concepts. Newton's book entitled "Hypothesis of Light" discussed that light was made up of particles or corpuscles that transported in all paths from the source. Newton argued that light waves, depending on its nature, could be twisted around hindrances. His concepts of light's nature of wave was against to the concept of light that traveled in straight path only. Nonetheless, he described the occurrence of light's diffraction by permitting particles of light to create a confined wave in an ether.

Theory of Newton helped to foresee the reflection of light, however if light hastened up when transported through a denser medium and due to greater gravitational pull, it is only then that refraction of light could be explained. Newton's concluded version of his theory was included in his "Opticks" which was published on 1704. The light's particle theory became influential during the 18th century because of his status.

Laplace argued that in the essence of particle theory of light, that light cannot escape from a body which is extremely huge; meaning it could be developed to a black hole. As the model for light has been recognized, Laplace withdrew his argument.

Polarization of light was qualitatively described by Newton for the first time using his particle theory. Then in 1810 a mathematical polarization theory of particle was designed by Etienne-Louis Malus. Later on 1812, Jean–Baptiste Biot evidently presented that this theory explained all occurrence of light polarization. It was polarization which considered as an evidence of light's particle theory.

Figure 2.1: Pierre Gassendi, supporter of Particle Theory.

Wave Theory

Robert Hooke (1635-1703) formulated a pulse theory and equated the scattering of light to the waves in bodies of water in order to explain the origin of colors. This was included in his work entitled "Micrographia" which was published on 1665. It was also him who proposed that the vibration of light might be perpendicular to the path of propagation. A mathematician named Christian Huygens (1629-1695) formulated the light's wave theory out of mathematical equation on 1678. "Treatise on Light" which was published on 1690 contained the work of Huygens. According to him Luminous ether was a medium responsible for the emission of light in all paths or direction.

Thomas Young noted that in wave theory, light waves similar to sound waves, it could obstruct or interfere each other. It was shown by Young through experiment conducted through diffraction and concluded that light deported as waves. He also suggested that through three colored receptors in the eyes, a person can see colors.

Leonhard Euler, a supporter of wave theory, believed that wave theory could undeniably expound diffraction in wave theory. If light acts as a transverse wave, the concept of polarization be illustrated easily; this was the idea of Augustin- Jean Fresnel which was hinted by André Marie Ampère.

Figure 2.2: Christian Huygens.

A year after, Fresnel studied on his own light's wave theory. Siméon Denis Poisson helped Fresnel through creating an argument that is convincingly favored wave theory that be downturned. Then in 1821, Fresnel formulated mathematical procedures that expounded by light's wave theory; it would be possible if the light is transverse in nature.

There was an identified weakness of wave theory, that in wave theory, light needs a medium in order to transmit.

Michelson-Morley's investigation caste extremely doubtful spell on the existence of luminiferous ether, which was proposed by Huygens.

In Newton's particle theory light transmit more rapidly in a denser medium, while Huygens, together with the supporters of wave theory implied, implicit the other side. These theories cannot be subjected to decision to which is accurate because the calculations of Speed of light is not precisely as well.

Leon Foucault, 1n 1850, formulated a precise measurement of the speed of light for the first time

And its outcome reinforced the wave theory; that made particle theory to be finally. Then in 20th century, its unexpected re-emerging partly was floated.

Electromagnetic Theory

Michael Faraday uncovered an effect called Faraday rotation. It is an effect described as when the light waves transmit endways in a magnetic ground with the aid of translucent dielectric, the undeviating polarized light rotates at the plane of polarization. This was the foremost proof that light was associated with magnetism. He guessed that light could be a form of vibration disseminating across the lines of magnetic ground. Then in 1847, Faraday suggested that light consisted of towering frequency of electromagnetic disturbance, light could proliferate even there was nonexistence of medium like ether.

James Clark Maxwell was inspired of Faraday's work, he also studied light and electromagnetic radiation. Maxwell uncovered that electromagnetic disturbance, which can propagate by itself, would transmit via space at a uniform speed. The calculations made by Maxwell was equal to the measurement of speed of light in the previous studies. Then in 1862, he stated in his book entitled "On Physical Lines of Force", that light was a type of an electromagnetic radiation. After several years, Maxwell published a book which consisted of his own equation called Maxwell Equations, a mathematical explanation of how electric and magnetic fields behaved and connected to each other.

Heinrich Hertz supported the theory of Maxwell, he further studied this theory by trial method by developing and noticing radio waves in his work place. These suspected radio waves acted equally the same with visible light; exhibit the characteristics of light namely reflection, refraction, diffraction and interference. Out of Maxwell's theory and Hertz' studies, modern radio, television, radar, wireless communication and electromagnetic imaging were developed.

Quantum Theory

Max Planck attempted to describe radiation called black body, and he proposed that light though considered as a wave, could acquire or reduce energy in limited amount associated with their frequency. Quanta, a Latin word which is derived from the phrase "how much" was the term used by Planck to call those mass of light energy.

Albert Einstein explained the photoelectric effect through the idea of quanta and proposed that these quanta really exist.

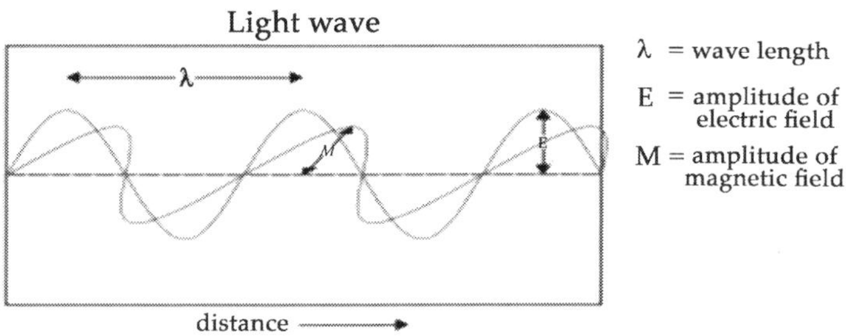

Figure 2.3: Shows the fluctuating components of light.

In 1923, Compton scattering was promulgated by Arthur Holly Compton, it explained how wavelength shifted when electrons emitted X- rays with low intensity. Compton scattering could be described by a theory of particle not of wave theory.

After three years, Gilbert N. Lewis used the term photon for the particles of light quanta.

Nowadays, Quantum mechanics, a modernized theory, visualized that light may be a particle and a wave, and an occurrence which is not a particle and not even a wave. Modern Physics pictures light as something that is difficult to comprehend. Physicists observed that the radio waves and X-rays in terms of their electromagnetic radiation, they have the tendency to behave like a natural wave in frequencies with lower intensity but a more natural particle when frequencies are higher, and yet, they both display all characteristics of wave and particle at the same time.

Visible light, which is at the middle rate of frequency, can be detected easily through experiments sand be described using wave or particle, or it could be both.

CHARACTERISTICS OF LIGHT

Light is the only electromagnetic radiation that human eye can perceived. It is composed of different wavelengths which can be pictured as different colors. Light that displayed a certain color is discharged on a thin ensemble of wavelengths.

Types of Light's Sources

- Incandescence light is described as the discharged light from hot matter.
- Luminescence light is described as the discharged of light when agitated electrons reduce its energy to lower level.

Light's Speed

The speed of light via vacuum or empty space is epitomized by the letter *c* from Latin word "celeritas" which means swiftness.

- The speed of light via vacuum or empty space is uniform in all locations.
- The speed of light via vacuum is constant at 299, 792, 458 m/s from the present description of meter.
- The speed of light via medium is constantly slower compared to that in a vacuum.
- The speed of light can be determined upon the type of medium which light transmit. The speed of light in all things with mass is constantly less than compared to the speed of light that travels in a vacuum.

Light wave's amplitude is associated with its intensity

- Intensity is described as the entire measure of power density of a light wave.
- Brightness is described as the intensity relatively of what the average human eye can be perceived.

The color of the light wave is related to its frequency.

Forms of Colors

a. Monochromatic light displayed one frequency only. Laser light is commendably monochromatic.

- Six colors which are associated with an ensemble of monochromatic light, Thesecolors namely red, orange, yellow, green, blue and violet; arranged according to their increasing frequency.

- Light is a common term referring to visible light; that inversely applicable to ultraviolet and infrared.

b. Polychromatic light displayed many different frequencies.

- Almost all light source is polychromatic.
- White light is an excellent example of polychromatic light.

Light wave's wavelength is reversely proportionate to its frequency.

- The longer the wavelengths of light the lower its frequency however, the shorter the wavelengths the greater its frequency.
- Light's wavelength range from 400 nanometer to 700 nanometer respectively the violet end to the red end of visible light spectrum.

Colors of the Light

The visible spectrum separates sunlight into several colors which are different from each other; this could happen when sunlight goes through a prism. The spectrum is composed of the colors of a rainbow namely red, orange, yellow, green, blue and violet. The most noticeable colors are red, green and blue; colors that the eyes are most sensitive to. The primary colors of light are red, green and blue. Combining these colors may result to different colors. For example when red and green light are united, the color is yellow; green plus blue gave birth to cyan, and red plus blue nay result to magenta.

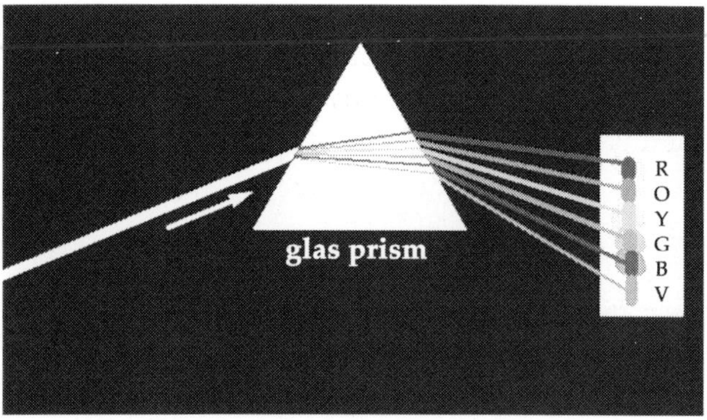

Figure 2.3: Colors of the light when travel through a prism.

THE HUMAN EYE

The human eyes are the most essential parts of a person. It is one of the body's five essential senses. The shapes, colors, and object's dimension can be interpreted by our eyes. This is the reason why eyes can be perceived as the most precious among the five senses.

Eyes can see in broad day light or even in a dim places, however they cannot see in inexistence of light. The light rays absorb by the eyes will be change into electrical signals then will be sent to the brain. The brain then will interpret it as visual images of objects.

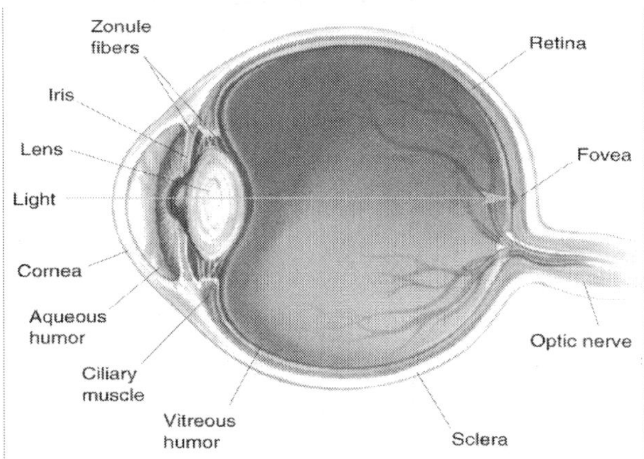

Figure 2.4: Anatomy of Human Eye.

The figure above shows the different parts of the eyes. The translucent outer membrane that forms the outer coating of the eyeball is called cornea. It functions as the eye's "window". The lens, a jelly like substance that consist of tiny fibers, causes strain to the light rays that enter the eye. The cornea then focus the light rays while lens is responsible for focusing the rays in its finer behavior to form a distinct image. The iris, the colored part of the eye located in front of the lens and composed of muscle fibers, determine the size of the pupil. Pupil then regulates the amount of light that pass through the eyes from the aqueous humour to the lens. It is an aperture which is controlled by the iris, for example in the absence of light the muscles of iris relax making the pupil to enlarge its opening in order for more light to enter and allow us to see in the dark. However, when the muscles of iris contract because of too much light, pupil then will become smaller or narrow to protect the cells of the eyes. To control the curvature of the lens, the ciliary muscle shrinks then enlarge. Vitreous humour can be found behind the lens, it is a cavity filled with liquid substance that gives nourishment to the parts of the eyes that are located in front and sustain the eye pressure. Retina, located just behind the vitreous humour, works as a screen where the images of the object are visualized. It is composed of millions of tiny nerve cells called rods and cones. These little nerve cells are very sensitive to the light, the cones are working when bright light exists and rods are responsible for the sight in places with dim conditions. Then from the retina, the light rays will enter a pathway called optic nerve, it will lead the rays to the processing area of the brain. Finally the brain will interpret the image of the object.

Facts about Human Eyes

These organs that are responsible for the sense of sight, perform very complicated function. So here is the procedure how the eyes work.

- When a person perceived an object, light mirrored it into the eyes.
- The light rays enter the eyes through the pupil. Movement of the pupil is dictated by the muscles of iris.
- Then, when the light has passed through the pupil, it will then enter the translucent lens of the eye. Refraction and focusing of the light rays occur on the retina.

- Sensitive cells called rods and cones that are present in the retina transform the light into electrical signals.
- Finally, optic nerve transmits these electrical signals to the brain which develops and interprets them as an image.

PRIMARY AND SECONDARY COLORS OF THE LIGHT

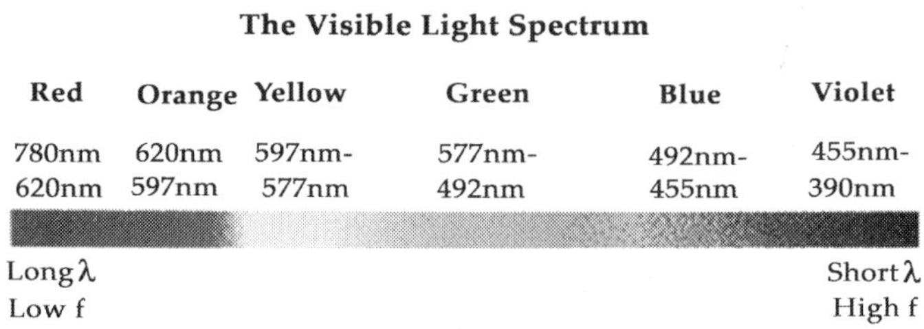

Figure 2.5: The Visible light Spectrum.

Primary colors of light can simplify the perception of colors. Visible light consists of different frequencies and represented through the visible light spectrum. The term white light is also referring to the entire colors of the light, the ROYGBV. The combinations of different range of frequency can produce white light, however, the combination of the three certain colors, which assumed to be separated widely in the spectrum can also produce white light. These three colors of light, combined with the same and correct intensity that can produce white light, are known as the primary colors. Red, green, and blue is the most distinct set of primary color. When these three frequencies summed up together with correct with correct intensity the result will be white light.

The combination of two or three of these primary colors can create a variety of colors. This is the reason why red, green and blue light that emit phosphors are used in television sets and computers.

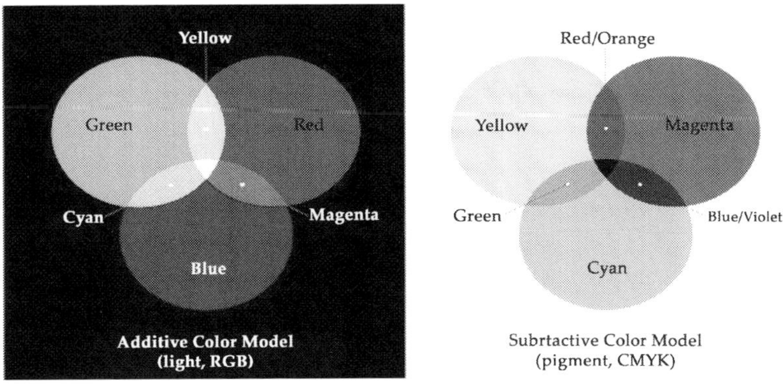

Figure 2.6: Shows the primary and secondary colors of light.

The figure above shows the primary colors in additive color model, that when combined with correct intensity, secondary colors are perceived. These secondary colors are the result of addition of colors of the primary colors. Yellow, cyan and magenta are the secondary colors that resulted from the combination of the primary colors. They are called secondary colors because they resulted from the combination of equal frequencies of two primary colors.

Then in subtractive color model, when yellow, magenta, and cyan are combined together the colors that you can get are the primary colors.

ADDITION AND SUBTRACTION OF LIGHT'S COLORS

Primary colors and secondary are discussed in the previous lesson, that white light is consisted of different colors represented by frequencies. These colors are red, orange, yellow, green, blue and violet. It is stated that there are three colors that can be combined to form white light. Red (R), Green (G) and Blue (B) are the primary colors; the mixture of two or more of these colors produced secondary colors. Yellow light is produced from the combination of Red and Green (R+G), Magenta is the result out of the combination of Red and Blue (R+B), lastly, when green is added to blue the result will be a cyan light.

Now, here is the discussion of addition and subtraction of the colors included in white light. A demonstration using light box can show the process of the addition of primary colors. This light box lights up a screen consisting of the three primary colors, the red (R), green (G) and Blue (B). The colors assume the shape of a circle. The overlapping of the two or more lighting in circles show the result of the combination of two primary colors.

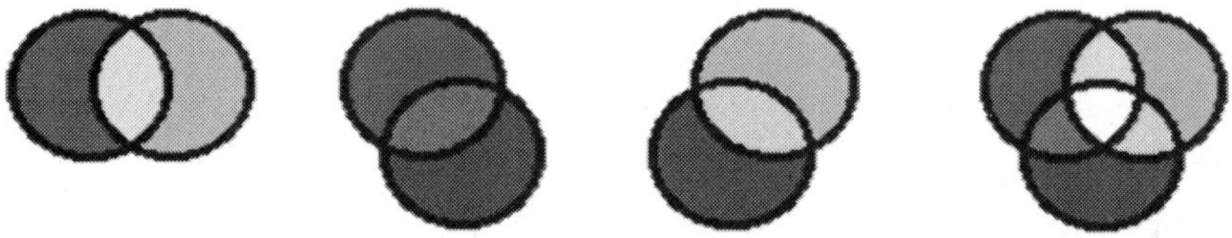

Figure 2.7: Shows the resulting colors when primary colors that are in circles overlap.

Rules in Addition of Colors

The following demonstrations exemplify that red light and green light when combined, yellow light will be produced. The combination of red light and blue light yields Magenta, when green light and blue light will be mixed together, the result is Cyan light. Lastly when red, green and blue light combined together, white light will be produced.

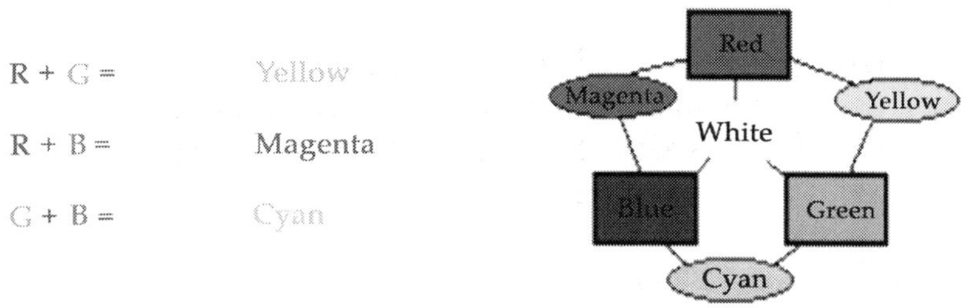

Figure 2.8: Shows equation and graphic presentation of addition of colors.

Color subtraction can be best demonstrated by using colored plastics and glass; they will be used as light's filters. These filters are attached to spotlights to produce the desired color. As the white light travel through a filter which is colored, some colors are removed. For instance, green filters permit mostly green, distinct blue and traces of yellow to pass through. The remaining colors of white light were absorbed.

In line with this, several filters must be used to remove all the colors of white light. Using a red filter placed then blue filter will be placed on top of the blue one. When light pass through the color red, a little orange and yellow will be permitted but the second filter which is blue in color will absorb all the colors that is why it appears black or no color are transmitted.

Paint mixing is one of the best way to illustrate the color subtraction. Green paint is produced out of the combination of yellow and blue paints. However, yellow and blue paints are impure colors. Tinctures of yellow paint will capture or deduct violet, indigo and blue from white light. Moreover, tincture of blue paint captured red orange and yellow colors as a result, both paints reflect green. The mixture of yellow and blue paints perceived to be green because the yellow or blue paint absorbed all other colors.

Eyes can perceive light, which is just a little part of the whole range radiations that transmit at the light's speed. It was in 18th century when Sir Isaac Newton conducted a famous experiment where in it demonstrated how light be split 7 colors namely red, orange, yellow, green, blue, indigo and violet. Indigo and violet are hard to distinguish from each other. These colors are very famous and they are also known as the colors of the rainbow. This refined and step by step investigation was conducted by Newton for the first time.

FUNDAMENTALS OF LIGHT

An electromagnetic radiation known as light is incomparable to anything in this world. Light cannot be considered as matter because it is not composed of particles of matter. From the concept of light being a wave or particle, analogies will be discussed, though not so ideal but it will relate the properties of light.

A light wave, as the waves in water, displays two characteristics, the distance between two successive crests and the distance between troughs. This distance refers to wavelength. Frequency is the other characteristic that is related to wavelength directly. For instance, the wavelength of the two waves measured 10 m apart, then 5 waves pass every second. The calculation for the speed is that10m x 1ms = 100 m/s. Therefore, wavelength multiply by frequency is equal to the speed of wave.

However, for light waves, its speed is constant; that one makes light special. Constantly frequency of light is associated with its wavelength.

Frequency and wavelength are the two terms in light wave that cannot be separated. When the wavelength is larger the frequency is lower and the smaller the wavelength the greater the frequency. Moreover the energy of wave is also related to its frequency. When the frequency is high which has shorter wavelength it means the energy it carries is high. When the waves have lower frequency which have longer wavelength, the energy it carries is low. Therefore, frequency, wavelength and energy are constantly related to each other.

How Fast the Light Travels?

Exactly 299, 792, 458 m/s is the speed of light through a vacuum. Meter is used as the SI units that define the speed of light. Radiations from all electromagnetic waves move completely the same in a vacuum. Throughout the history, a number of physicist have tried to calculate the exact speed of light. In 17th century, Galileo made an effort to measure light's speed. However in 1676, a Danish physicist named Ole Rømer conducted an experiment to measure the speed of light. He studied the movement of Jupiter and Io, one of the Jupiter's moons. He observed inconsistencies in the actual time of Io's orbit, he premeditated that light consumes about twenty two minutes to navigate the diameter of the Earth's orbit. Unfortunately, the size of the diameter of the Earth's orbit was not determined at that time. If only he knew the diameter of the Earth's orbit, he would have described its speed to be 227, 000, 000 m/s.

A more accurate calculation of the speed was conducted by Hippolyte Fizeau. The experiment was performed in Europe on 1849. He positioned a beam light at mirror few kilometers at a distance. A wheel with cogs was stationed at the track of the light beam as it transmitted from the source, directed to the mirror and then moved back to its original position. Fizeau learned that at a certain amount of rotation, the beam would navigate through one gap in the wheel then in its way back another gap would pass through. If he had known the distance to the mirror, the quantity of the wheel's teeth and the amount of rotation, the speed of light that Fizeau would have calculated was 313, 000, 000 m/s.

In 1862, an experiment was carried out by Leon Foucalt; he used mirrors that are rotating to acquire a value of 298, 000, 000 m/s. Albert Michelson polished the methods done by Foucalt through conducting experiments on the light's speed until he died in 1931. Later in 1926,

Michelson used mirrors that are rotating and made some improvements on them. These mirrors calculated the time it acquired in order to yield a round expedition from Mount Wilson to Mount San Antonio located in California. Through this experiment an accurate measurement of light speed was determined. The speed of light is measured 299, 796, 000 m/s.

In different translucent substances that are composed of conventional matter, the operative velocity is less compared in vacuum. For instance, the speed of light in water is approximately three fourths of in a vacuum.

Units and Measurements used in Light

Radiometry and photometry are the two foremost alternative sets of units to be used in measuring light. Radiometry composes of measurements of light power at all wavelengths. However, photometry calculates light with wavelength subjective to the concept of consistent model of human brightness perception. Photometry is beneficial to compute brightness proposed for human use. Here is the summarized SI units for both radiometry and photometry.

SI Radiometry Units

Table 1

Quantity		Unit		Dimension	Notes
Name	Symbol[nb 1]	Name	Symbol	Symbol	
Radiant energy	Q_e[nb 2]	joule	J	$\mathbf{M}\cdot\mathbf{L}^{2}\cdot\mathbf{T}^{-2}$	Energy of electromagnetic radiation.
Radiant energy density	w_e	joule per cubic metre	J/m^3	$\mathbf{M}\cdot\mathbf{L}^{-1}\cdot\mathbf{T}^{-2}$	Radiant energy per unit volume.
Radiant flux	Φ_e[nb 2]	watt	W *or* J/s	$\mathbf{M}\cdot\mathbf{L}^{2}\cdot\mathbf{T}^{-3}$	Radiant energy emitted, reflected, transmitted or received, per unit time. This is sometimes also called "radiant power".
Spectral flux	$\Phi_{e,\nu}$[nb 3] or $\Phi_{e,\lambda}$[nb 4]	watt per hertz *or* watt per metre	W/Hz *or* W/m	$\mathbf{M}\cdot\mathbf{L}^{2}\cdot\mathbf{T}^{-2}$ *or* $\mathbf{M}\cdot\mathbf{L}\cdot\mathbf{T}^{-3}$	Radiant flux per unit frequency or wavelength. The latter is commonly measured in W·sr^{-1}·m^{-2}·nm^{-1}.

Radiant intensity	$I_{e,\Omega}$ [nb 5]	watt per steradian	W/sr	$\mathbf{M{\cdot}L^{2}{\cdot}T^{-3}}$	Radiant flux emitted, reflected, transmitted or received, per unit solid angle. This is a *directional* quantity.
Spectral intensity	$I_{e,\Omega,\nu}$ [nb 3] or $I_{e,\Omega,\lambda}$ [nb 4]	watt per steradian per hertz *or* watt per steradian per metre	$W{\cdot}sr^{-1}{\cdot}Hz^{-1}$ *or* $W{\cdot}sr^{-1}{\cdot}m^{-1}$	$\mathbf{M{\cdot}L^{2}{\cdot}T^{-2}}$ *or* $\mathbf{M{\cdot}L{\cdot}T^{-3}}$	Radiant intensity per unit frequency or wavelength. The latter is commonly measured in $W{\cdot}sr^{-1}{\cdot}m^{-2}{\cdot}nm^{-1}$. This is a *directional* quantity.
Radiance	$L_{e,\Omega}$ [nb 5]	watt per steradian per square metre	$W{\cdot}sr^{-1}{\cdot}m^{-2}$	$\mathbf{M{\cdot}T^{-3}}$	Radiant flux emitted, reflected, transmitted or received by a *surface*, per unit solid angle per unit projected area. This is a *directional* quantity. This is sometimes also confusingly called "intensity".
Spectral radiance	$L_{e,\Omega,\nu}$ [nb 3] or $L_{e,\Omega,\lambda}$ [nb 4]	watt per steradian per square metre per hertz *or* watt per steradian per square metre, per metre	$W{\cdot}sr^{-1}{\cdot}m^{-2}{\cdot}Hz^{-1}$ *or* $W{\cdot}sr^{-1}{\cdot}m^{-3}$	$\mathbf{M{\cdot}T^{-2}}$ *or* $\mathbf{M{\cdot}L^{-1}{\cdot}T^{-3}}$	Radiance of a *surface* per unit frequency or wavelength. The latter is commonly measured in $W{\cdot}sr^{-1}{\cdot}m^{-2}{\cdot}nm^{-1}$. This is a *directional*
Irradiance	E_{e} [nb 2]	watt per square metre	W/m^{2}	$\mathbf{M{\cdot}T^{-3}}$	Radiant flux *received* by a *surface* per unit area. This is sometimes also confusingly called "intensity".
Spectral irradiance	$E_{e,\nu}$ [nb 3] or $E_{e,\lambda}$ [nb 4]	watt per square metre per hertz *or* watt per square metre, per metre	$W{\cdot}m^{-2}{\cdot}Hz^{-1}$ *or* W/m^{3}	$\mathbf{M{\cdot}T^{-2}}$ *or* $\mathbf{M{\cdot}L^{-1}{\cdot}T^{-3}}$	Irradiance of a *surface* per unit frequency or wavelength. The terms spectral flux density or more confusingly "spectral intensity" are also used. Non-SI units of spectral irradiance include Jansky = 10^{-26} $W{\cdot}m^{-2}{\cdot}Hz^{-1}$ and solar flux unit (1SFU = 10^{-22} $W{\cdot}m^{-2}{\cdot}Hz^{-1}$).
Radiosity	J_{e} [nb 2]	watt per square metre	W/m^{2}	$\mathbf{M{\cdot}T^{-3}}$	Radiant flux *leaving* (emitted, reflected and transmitted by) a *surface* per unit area. This is sometimes also confusingly called "intensity".
Spectral radiosity	$J_{e,\nu}$ [nb 3] or $J_{e,\lambda}$ [nb 4]	watt per square metre per hertz *or* watt per square metre, per metre	$W{\cdot}m^{-2}{\cdot}Hz^{-1}$ *or* W/m^{3}	$\mathbf{M{\cdot}T^{-2}}$ *or* $\mathbf{M{\cdot}L^{-1}{\cdot}T^{-3}}$	Radiosity of a *surface* per unit frequency or wavelength. The latter is commonly measured in $W{\cdot}m^{-2}{\cdot}nm^{-1}$. This is sometimes also confusingly called "spectral intensity".

Radiant exitance	M_e [nb 2]	watt per square metre	W/m²	$\mathbf{M}\cdot\mathbf{T}^{-3}$	Radiant flux *emitted* by a *surface* per unit area. This is the emitted component of radiosity. "Radiant emittance" is an old term for this quantity. This is sometimes also confusingly called "intensity".
Spectral exitance	$M_{e,v}$ [nb 3] or $M_{e,\lambda}$ [nb 4]	watt per square metre per hertz *or* watt per square metre, per metre	$W\cdot m^{-2}\cdot Hz^{-1}$ *or* W/m³	$\mathbf{M}\cdot\mathbf{T}^{-2}$ *or* $\mathbf{M}\cdot\mathbf{L}^{-1}\cdot\mathbf{T}^{-3}$	Radiant exitance of a *surface* per unit frequency or wavelength. The latter is commonly measured in $W\cdot m^{-2}\cdot nm^{-1}$. "Spectral emittance" is an old term for this quantity. This is sometimes also confusingly called "spectral intensity".
Radiant exposure	H_e	joule per square metre	J/m²	$\mathbf{M}\cdot\mathbf{T}^{-2}$	Radiant energy received by a *surface* per unit area, or equivalently irradiance of a *surface* integrated over time of irradiation. This is sometimes also called "radiant fluence".
Spectral exposure	$H_{e,v}$ [nb 3] or $H_{e,\lambda}$ [nb 4]	joule per square metre per hertz *or* joule per square metre, per metre	$J\cdot m^{-2}\cdot Hz^{-1}$ *or* J/m³	$\mathbf{M}\cdot\mathbf{T}^{-1}$ *or* $\mathbf{M}\cdot\mathbf{L}^{-1}\cdot\mathbf{T}^{-2}$	Radiant exposure of a *surface* per unit frequency or wavelength. The latter is commonly measured in $J\cdot m^{-2}\cdot nm^{-1}$. This is sometimes also called "spectral fluence".
Hemispherical emissivity	ε		1		Radiant exitance of a *surface*, divided by that of a *black body* at the same temperature as that surface.
Spectral hemispherical emissivity	ε_v or ε_λ		1		Spectral exitance of a *surface*, divided by that of a *black body* at the same temperature as that surface.
Directional emissivity	ε_Ω		1		Radiance *emitted* by a *surface*, divided by that emitted by a *black body* at the same temperature as that surface.
Spectral directional emissivity	$\varepsilon_{\Omega,v}$ or $\varepsilon_{\Omega,\lambda}$		1		Spectral radiance *emitted* by a *surface*, divided by that of a *black body* at the same temperature as that surface.

Hemispherical absorptance	A	1	Radiant flux *absorbed* by a *surface*, divided by that received by that surface. This should not be confused with "absorbance".
Spectral hemispherical absorptance	A_ν or A_λ	1	Spectral flux *absorbed* by a *surface*, divided by that received by that surface. This should not be confused with "spectral absorbance".
Directional absorptance	A_Ω	1	Radiance *absorbed* by a *surface*, divided by the radiance incident onto that surface. This should not be confused with "absorbance".
Spectral directional absorptance	$A_{\Omega,\nu}$ or $A_{\Omega,\lambda}$	1	Spectral radiance *absorbed* by a *surface*, divided by the spectral radiance incident onto that surface. This should not be confused with "spectral absorbance".
Hemispherical reflectance	R	1	Radiant flux *reflected* by a *surface*, divided by that received by that surface.
Spectral hemispherical reflectance	R_ν or R_λ	1	Spectral flux *reflected* by a *surface*, divided by that received by that surface.
Directional reflectance	R_Ω	1	Radiance *reflected* by a *surface*, divided by that received by that surface.
Spectral directional reflectance	$R_{\Omega,\nu}$ or $R_{\Omega,\lambda}$	1	Spectral radiance *reflected* by a *surface*, divided by that received by that surface.

Hemispherical transmittance	T		1	Radiant flux *transmitted* by a *surface*, divided by that received by that surface.
Spectral hemispherical transmittance	T_{ν} or T_{λ}		1	Spectral flux *transmitted* by a *surface*, divided by that received by that surface.
Directional transmittance	T_{Ω}		1	Radiance *transmitted* by a *surface*, divided by that received by that surface.
Spectral directional transmittance	$T_{\Omega,\nu}$ or $T_{\Omega,\lambda}$		1	Spectral radiance *transmitted* by a *surface*, divided by that received by that surface.
Hemispherical attenuation coefficient	μ	reciprocal metre m^{-1}	L^{-1}	Radiant flux *absorbed* and *scattered* by a *volume* per unit length, divided by that received by that volume.
Spectral hemispherical attenuation coefficient	μ_{ν} or μ_{λ}	reciprocal metre m^{-1}	L^{-1}	Spectral radiant flux *absorbed* and *scattered* by a *volume* per unit length, divided by that received by that volume.
Directional attenuation coefficient	μ_{Ω}	reciprocal metre m^{-1}	L^{-1}	Radiance *absorbed* and *scattered* by a *volume* per unit length, divided by that received by that volume.
Spectral directional attenuation coefficient	$\mu_{\Omega,\nu}$ or $\mu_{\Omega,\lambda}$	reciprocal metre m^{-1}	L^{-1}	Spectral radiance *absorbed* and *scattered* by a *volume* per unit length, divided by that received by that volume.

(Source: www.google.com.)

SI Photometry Units

Table 2

Quantity Name	Symbol [nb 1]	Unit Name	Symbol	Dimension Symbol	Notes
Luminous energy	Q_v [nb 2]	lumen second	lm·s	T·J [nb 3]	Units are sometimes called *talbots*.
Luminous flux / luminous power	Φ_v [nb 2]	lumen (= cd·sr)	lm	J [nb 3]	Luminous energy per unit time.
Luminous intensity	I_v	candela (= lm/sr)	cd	J [nb 3]	Luminous power per unit solid angle.
Luminance	L_v	candela per square metre	cd/m²	L^{-2}·J	Luminous power per unit solid angle per unit *projected* source area. Units are sometimes called *nits*.
Illuminance	E_v	lux (= lm/m²)	lx	L^{-2}·J	Luminous power *incident* on a surface.
Luminous exitance / luminous emittance	M_v	lux	lx	L^{-2}·J	Luminous power *emitted* from a surface.
Luminous exposure	H_v	lux second	lx·s	L^{-2}·T·J	
Luminous energy density	ω_v	lumen second per cubic metre	lm·s·m^{-3}	L^{-3}·T·J	
Luminous efficacy	η [nb 2]	lumen per watt	lm/W	M^{-1}·L^{-2}·T^3·J	Ratio of luminous flux to radiant flux or power consumption, depending on context.
Luminous efficiency / luminous coefficient	V			1	

(Source: www.google.com)

WAVE INTERFERENCE OF A LIGHT

When two waves encountered each other while transmitting along a similar medium, this occurrence is called wave interference. This interference is the reason why medium assumed a shape that is the outcome from the net effect of the two waves against the elements of the medium. Wave interference may be classified as constructive or destructive in origin. Constructive interference transpires at every position in a medium where the displacement of the two waves, that are interfering, travels in the same path. For instance, if a constant given in time and position in a wave encounter the highest peak of the first wave encountered the highest peak of the second wave, then the result is "super crest", correspondingly, two troughs that interfere yield a "super trough".

When the two intrusive waves have displaced in the reverse path, it is called as destructive interference. The intrusion happened in a crest with a trough is a best example of destructive interference. Destructive interference is inclined to reduce the subsequent displacement of the medium.

There was a crucial moment in the history of the argument about the natural concept of light in the beginning of the nineteenth century. When light from two causes of waves transmit via similar medium resulted to an interference pattern as shown by Thomas Young. Young's experiments consisted of few steps and explained the water wave's interference which was created from two points.

Two-point Source interference Patterns

It is a pattern of two groups of periodic and concentric waves displaying similar frequency that creates a fascinating arrangement in a ripple tank. If an object emerge and sink in the water rapidly, circles with the same center will be propagated on the water. When two objects rapidly emerge and sink with similar frequency on different points, therefore two sets of circular waves, having the same center, will be propagated on the surface of the water. The interference of these circles or concentric waves will be observed as they transport crossways the surface of the water.

Is two-point source interference be produced by light?

When a light beneficially obstruct, like a crest collides to a crest or trough collides to a trough, the two waves move to strengthen one another and yield a super light wave. However, when a light destructively obstructs, like a crest collides with a trough, the two waves move to extinguish each other and no light will be yielded. Therefore, the two-point source interference pattern is composed of repeated series of antinodal lines and nodal lines. For light waves, bright lines correspond to the antinodal lines and dark lines are equal to that nodal lines. Then if that pattern of interference be developed two light sources and applied on a screen, the result will be repeated series of bright and dark strips on the screen. In addition to that, the pattern used in that occurrence was antinodal lines therefore the central strip projected on the screen will be bright strip of light. Thomas Young proved that light can propagate two-point source interference pattern. This study was conducted in 1801.

For the light to produce that pattern, the light that must be utilize was monochromatic light. Monochromatic light is composed of single color only. To use this light, two bases will radiate with equal frequency, that the highest peak of the first wave must be propagated with equal accurate time as the highest peak of the second wave. In order for Young to complete his work, he used one light source and applied the light into two pinpricks. Through these pinpricks the light will diffract and the design will be shown on a screen. Due to single source of light, the series of two waves that originate from the pinpricks will be at the same point with each other. As a result, a series of repeated dark and bright strips was produced with the use of monochromatic light and pinpricks.

VISIBLE LIGHT SPECTRUM

The visible region which is characterized by a very thin strips of wavelengths that can be found at the right of the infrared section and at the left of the ultra violet section. Human eyes are sensitive to this thin band of wavelengths. The term visible light spectrum was used since human can see it. This is the only electromagnetic wave that can excite the eye's retina and consists of small range of frequency of radiation compare to the other types of the electromagnetic waves that are composed of massive frequencies of electromagnetic radiation.

The section of visible light is composed of a range of wavelengths that measured nearly 700 nanometers to almost 400 nm (7×10^{-7} nm to 4×10^{-7} nm). This thin strips of visible light is famously known as the ROYGBIV.

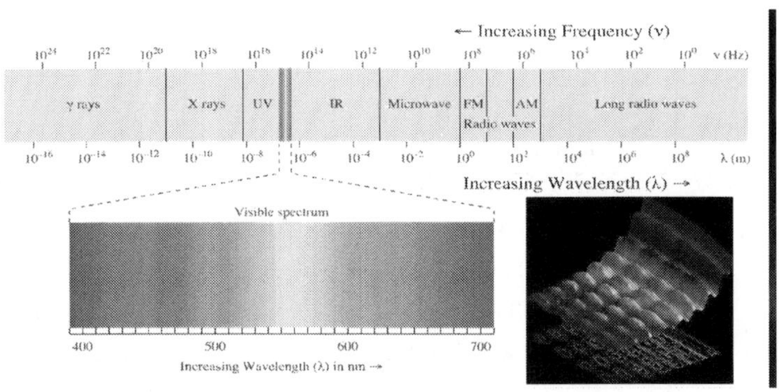

Figure 2.9: The visible light spectrum from the electromagnetic spectrum.

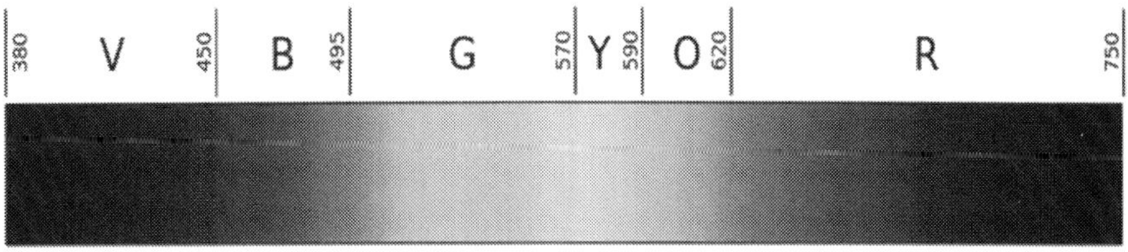

Figure 2.10: The wavelengths of the Visible Light Spectrum.

Each particular color is represented by an individual wavelength within the spectrum. The reason why when a certain wavelength hits the retina of the eye, that particular color will be perceived by the eyes. It was Newton who showed that a light transmit through a prism will be projected into separated wavelengths and display different colors that the visible light is composed of. Dispersion is the process of separating visible light into various colors. Each color is characterized by a specific wavelength and these diverse wavelengths will be bent accordingly to the varying amount as entering the prism. A prism is the passage way for the visible light to be dispersed. The following colors are propagated when the visible light was dispersed; red, orange, yellow, green, blue and violet. The other name for visible light actually came from these colors, R is for red, O is for orange, Y is for yellow, G is for green, B is for blue and V is for violet; that name pertains to ROY G. BIV. Essentially, letter I stands for Indigo but incidentally it is not observed in the spectrum, it is usually attached to the acronym so that it functions as the vowel in ROY's surname. The longer wavelengths in the spectrum belong to red and the shorter wavelengths belong to violet.

White color is perceived when all the wavelengths of the visible light assault the eyes. The impression of white is the outcome of the combination of all the colors in the visible spectrum. Therefore, ROYGBIV the combination of all the colors in the spectrum is also known as white light. Theoretically, white cannot be described as a color but not that logically there is a wavelength that corresponds to the behavior of white but white can be observed as the mixture of all the wavelengths or colors of the visible light. Then it is concluded that all the colors of the visible light lead to the propagation of white light, therefore there is no wavelength that will be able to produce black color. Just like white, technically speaking black cannot be perceived as color, it is simply the nonappearance of the wavelengths of the visible light. So at night time, you only see darkness because of the absence of the visible light spectrum's wavelengths that are assaulting the eyes.

ABSORPTION, REFLECTION AND TRANSMISSION OF LIGHT

Visible light is composed of constant spectrum of wavelengths and frequencies. There are many possibilities could happen when a light wave that has single frequency, assault an object. It could be absorbed, reflected and transmitted by the object. What will happen if there are many frequencies that strike the object? In the occurrence of this situation, the object has the tendency to absorb, reflect and transmit particularly, for example, an object might project yellow light then all other colors will be absorbed. Therefore the interaction between the light and the object depends on the frequency of the wavelengths of light and the object's atomic component.

CHAPTER 3

ELECTROMAGNETIC WAVES: WIDER SCOPE

INTRODUCTION

Energy conduction through a vacuum or require no medium define electromagnetic wave. It is caused by fluctuations of electric and magnetic grounds. An electromagnetic wave conveyed its energy through an empty space or vacuum at a constant speed of 300 x 10^8 m/s. The transfer of energy of electromagnetic waves includes absorption and reemission of the wave energy by the atoms of the material. When the energy of an electromagnetic wave intrudes the atom, it absorbs the energy.

What is electromagnetic radiation? Its meaning is slightly difficult to comprehend, however people interacts with electromagnetic radiation in almost all aspects of their lives. Electromagnetic radiation is propagated through the vibration of the magnetic grounds.

Electromagnetic waves are utilized to send various size of wavelength of radio waves and television signals. Radio waves, microwaves, visible light, ultraviolet waves, X-ray and gamma rays are being transported by the electromagnetic waves. The electromagnetic spectrum has different sections which play a vital role in the lives of people living on Earth and in the field of industry which involves technology in communication.

TYPES OF ELECTROMAGNETIC WAVES

Radio Waves

James Clark Maxwell predicted radio waves through his mathematical equations in 1867. He observed this wavelike properties which is the same with light and resemble from electrical and magnetic observations. He characterized radio waves and light waves as electromagnetic waves that transmitted in space, emitted by an electric particle while accelerating, and this is the inclusion of Maxwell's equations.

Heinrich Hertz, 1887, confirmed the truth about Maxwell's theory by doing an experiment in his laboratory, he propagated radio waves and showed that just like light, radio waves exhibit standing waves, refraction, polarization and diffraction. It was Guglielmo Marconi who established the first radio transmitters and receivers that were used for communication in the middle of 1890s.

Techniques used by Radio system in Propagating Radio waves for Communication

Line of Sight

Pertains to radio waves that transmit along straight line from the transferring to the receiving antenna. Radio waves at lower frequencies can penetrate through establishments and other barriers, therefore it does not require an unoccupied track. The route of sight propagation on the Earth's surface is confined by the graphic vanishing point to approximately 40 miles. Cellphones, cordless phones, wireless networks, FM and television and radar used this method for the propagation of radio waves or transmitted signals. Dish antennas are used to transfer microwave beams; telephone and television signals were linked by radio relay to the distance as far as the visual of horizon. Ground stations are used to connect to satellites and spacecraft from Earth, travelling through billions of miles.

Ground waves

Radio waves that belongs to lower frequencies can be twisted above the mounds of earth like hills and mountains, and produced further than that of vanishing point of sight. They travel like the surface waves that track the outline of the Earth. Medium-wave and long-wave radio station are allowed by ground waves to cover places as far as the vanishing point of sight or the horizon, beyond hundreds of miles. Ground waves decreased its power and its wave fronts twist or angle as the surface of the Earth absorbs its energy.

Tropospheric propagation

Radio waves can transmit further compared to the line of horizon in line to refraction in the lowermost layer of Earth's atmosphere at 20 km. It happens when air refracts with temperature and pressure. Through tropospheric vessel, radio waves can transmit within the range of 500 to 1000 km.

Sky waves

Radio waves reflecting ions at medium wave and shortwave in the atmospheric layer called ionosphere, located above the Earth at 150-500 km. Radio waves were sent at an angle to the sky and returned to Earth beyond the line of sight. The propagation of sky waves varies and depends on conditions of the atmosphere.

Radio waves are utilized for telecommunications of information. These information are disseminated on wave carriers as amplitude, modulation or AM and frequency amplitude or FM. A frequency band which thickness is equal to the density of information is involved in the transmission. No single-frequency electromagnetic wave is being transmitted. The thickness of the band is about 10,000 Hz for telephone, 20,000 Hz for very good quality of recorded sound, and five megahertz for television that has high definition. Radio wave was confined to a lower frequency just near 10, 000 Hz because of the decreasing efficiency of generating electromagnetic waves.

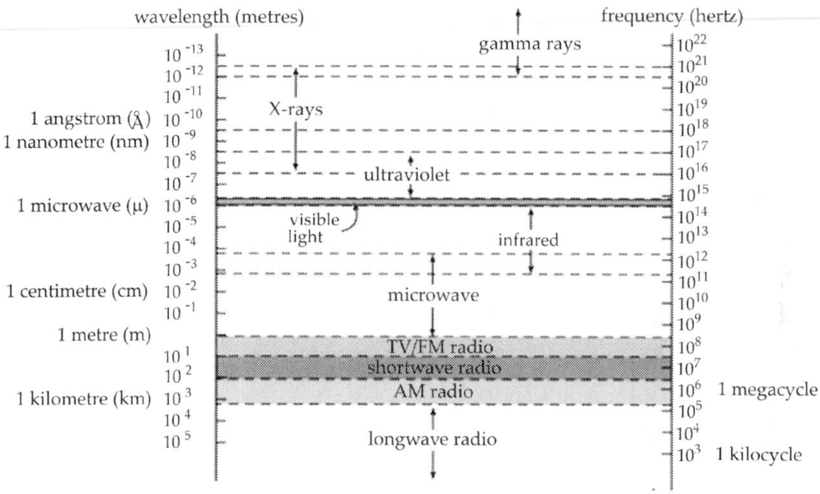

Figure 3.1: Electromagnetic spectrum.

Electromagnetic disturbance transmits in empty space freely and in straight lines, therefore the inflection of Earth confines the distance of the observer's eye to a distant point from the top of tower which is 100 m or about 30 km. The unpredicted accomplishment of Marconi in communicating messages in the range of more than 1,200 miles had opened the path to the discovery of the ionosphere or Kennelly-Heaviside layer. The layer of atmosphere which is approximately 300 km thick, lies about 100 km above the Earth's surface. The atmosphere in this region is somewhat ionized by the sun's ultraviolet light. The atmosphere is partially converted into ions by the sun's ultraviolet light, excite ample amount of electrons and ions that affect radio waves. The ionization of arranged ionosphere depend from day to night and from summertime to wintertime, which is because of the influence of the sun. Ionosphere twisted or even reproduced the radio waves being sent through antennas in particular location. Radio telecommunication became possible because of the ionosphere's repeatedly movement in bouncing off and reflecting radio waves.

Ground wave enabled long distance communication more vividly. The surface of the Earth especially above water was closely followed by the form of electromagnetic wave; an outcome of the wave's contact with earthly surface. The frequency of the wave dictate the ground wave range and the bending and reproduction of the sky wave through ionosphere. Forty Mega Hertz, under normal condition of ionosphere, is the uppermost frequency that radio wave had, that can be transmitted from ionosphere under normal circumstances. Television occurrences must be higher than 40 MHz in order for the conveyed signal to be taken by the ionosphere. This is the reason why transmitters of television are placed on high places such as towers and hilltops.

Radio waves may be interrupted by tall establishments and other hindrances as it transported from transmitter to the antenna which receives the signals. When some parts of the waves reflected upon departing to the antenna and delay the admission of the transmitted signals then disturbances developed. Woods, bricks and concrete walls are some of the conducting materials that can be infiltrated by the radio waves. However, water and metals which are good conductors of electricity cannot be infiltrated by radio waves. The observations conducted by scientists doing radio astronomy be possibly done using ground-based telescopes because radio waves from outer space can pierce through the atmosphere of the Earth. Wires and coaxial cables are used to restrict the perception of the radio waves into limited places. This method reduces energy loss and disturbance of the waves whenever transmitted from one place to another. For radio and television, elcctromagnetic energy is transmitted through unguided or wireless communication because receivers' locations are indefinite in number. Coaxial cable system transmit radiation to their users straightly from the broadcasting stations or from an antenna of specified community. This kind of guide transmission delivers signals that are high in quality.

Microwaves

The microwave region is ranging from 1, 000 to 300, 000 MHz which is equivalent to 30 cm to 1mm wavelength. This electromagnetic radiation transmits signals to radar, televisions and satellites. Generally microwaves perform more complex function; not just for heating left over food to be consumed again. They display higher frequencies consisting of cycles that are taking place in every second. Microwaves were first discovered and studied in 1886 but its useful presentation took many years to be applied in klystron and magnetron. Microwaves are the main transporters of transmission that are of high speed data among stations such as stations on Earth, ground-based stations, satellites and space probes. Satellites which are concurrent or existing at the same time are located 36, 000 km above the Earth's atmosphere, and are used to mark the comprised wideband of communication such as televisions and telephones.

Parabolic dish antennas are used by microwave transmitters and receivers of microwave. Microwave beams were produced by these antennas which spread out at angle proportionate to the percentage of integral wavelength to the distance crossways of the dish. Short pulses of microwaves comprised radar beams. In defining the distance of a ship, the time threads on the pulse to transmit and then reflected again on radar antenna. In order to obtain the measurement of the speed of the object, utilize the reflected wave. Pulse's change in frequency is caused by Doppler Effect. Guiding airplanes and determining over speeding motorists are the function by which microwave radar are comprised of. Microwaves are also useful in charting meteorological disturbances and weather forecasting since it can pierce through the cloud of smoke and separated by water droplets.

As stated earlier, microwave plays a vital role in terms of cooking where food needs to be heated and cooked using this method. The stored water and fats from food absorb the propagated heat that happens inside. Unlike radio waves at lower frequency which do not diffract over the mount of soil, microwave transmit through line of sight and track the surface of the Earth just like ground waves, then replicate by the ionosphere. This is the reason why communication through microwave are confined to 64 km.

Modern technology uses microwaves in communication links, wireless networks, microwave radio delay networks, radar, communication in satellite and space craft, medical and cancer treatment.

High-powered sources of microwaves use specified vacuum pipes to propagate microwaves. Using the electron's airborne motion which is controlled by magnetic fields, these devices function depending on rules from vacuum tubes which are at lower frequency. The devices of high-powered microwaves function mainly based on the airborne motion of electrons which travel on them than using a flowing stream of electrons. Sources of microwave with low power capacity utilized made of solid devices. Field effect transistors, tunnel diodes, Gun diodes and IMPHTT diodes are the devices that are used by low-power microwaves to propagate. If light has a device called laser that increases the propagation of light, microwave has its own version called Maser, and it is used to intensify microwaves.

Black body radiations are low level microwave was discharged by all somewhat hot objects. In order to determine the temperature of an object or topography scientists used microwave radiometers.

Heating Application

Microwave ovens transformed into kitchen equipment in Western part of the world in 1770s. It uses microwave radiation with a frequency about 2.45 GHz, causing the heating of nonconductor of direct electric current. Drying and preserving of products are the industrial processes that microwave heating is utilized for.

Microwave in electromagnetic spectrum

Name	Wavelength	Frequency (Hz)	Photon Energy (eV)	Range width (Bel)
Gamma ray	<0.02nm	>15 EHz	>62.1 keV	infinite
X-ray	0.01 nm- 10 nm	30 EHz- 30 PHz	124 KeV-124 eV	3
Ultraviolet	10nm- 400 nm	30 PHz- 750 THz	124 eV- 3 eV	1.6
Visible light	390 nm -750 nm	770 THz- 400 THz	3.2 eV- 1.7 eV	0.3
Infrared	750 nm- 1mm	400 THz- 300 GHz	1.7 eV-1.24 meV	3.1
Microwave	1mm- 1m	300 GHz- 300 MHz	1.24 meV- 1.24 μeV	3
Radio	1mm- 100, 000 km	300 GHz- 3 Hz	1.24 meV-12.4 feV	8

In this Electromagnetic spectrum, microwave resides in between infrared and radio waves.

High gain antennas produced narrow beams of microwaves with short wavelengths from a half meter to five meters in diameter. Consequently communication links known as point- to- point and radar use beam microwaves. These narrow beams permits frequency reuse by receivers that are nearer. Microwave frequencies utilized directive antennas such as parabolic dish antenna, however some other antennas are used such as horn antenna, slot antennas and dielectric lens. Consumer device which is increasingly used by consumer is the flat micro strip antennas. Phased array is another directive antenna applied at microwave frequency, it is being controlled by a computer which proliferate electronically stirred beams that are being sent in different directions. Wireless devices and Wi-Fi routers for wireless LAN, antenna used by cellphones such as inverted F antenna, and dipole antennas are utilized. It is where omnidirectional antennas are used mandatory. Coaxial and parallel wire lines at microwave frequencies used to convey lower frequency radio waves. As the microwaves frequency transmit in transmission lines, it is experiencing too much power losses, in this case wave guides metal pipe carried microwaves when low reduction is necessary.

Living tissues were damaged as they were exposed to the heating effect of microwaves where in the temperature of the tissue surpass 43^0 C. Consequently, having contact to strong microwaves is harmful especially above 20 milliwatts of power per square centimeters. Cataracts are the results of repeatedly, contact of lens of the human eye to the waves with precise range of 300 MHz About 19-20 milliwatts per square meters was needed in order to produce heating that can adversely influential to the brain's electron and chemical balance and the developing babies.

Infrared Waves

Infrared radiation is an electromagnetic wave with wavelengths that are longer compared to visible light, though invisible it is occasionally called infrared light. In visible spectrum, it is located at the red section at 200 nm to 1, 000, 000 nm. Infrared are mostly heat radiation expelled by objects at room temperature. Infrared discharged radiant energy and characterized as wave and particle.

Sir William Herschel, an astronomer, who discovered infrared in 1800. He discovered an invisible radiation through its effect on a thermometer. It is located in the spectrum's region where energy is lower compared to red light. According to the studies, the energy of the sun arrived to Earth in the form of infrared. The energy that arrived on Earth found out to be more than half of the total energy emitted by the sun. The equilibrium in the absorbed and emitted infrared has an advantageous effect on the climate of the Earth. When the molecules charge their rotational vibrational movements, infrared radiation is expelled or sustained. A change in the dipole moment is caused when infrared stimulates the mode of vibrations in molecules, its range of frequency will be helpful to study the energy range of the molecules in proper symmetry. The photos absorbed and transmitted included in a scope of infrared are examined through Infrared spectroscopy. Scientific, medical and industrial applications used infrared radiation. The use of right vision equipment, which is irradiated by infrared, permitted observers without being noticed. Infrared astronomy utilizes devices with sensor equipped such as telescope to pierce through regions in space that are dusty regions called molecular clouds, planets are also detected and the objects with red radiance will be observed. Detecting heat loss, studying altering blood flow through the skin and observing an excessive heating of apparatuses operated by electricity, are the uses for the thermal imaging cameras using infrared. Military operations use thermal-infrared imaging. It is very beneficial to then in target acquisition, surveillance, night vision for night activities and tracking of people and vehicles. Normal body temperature of a person range at wavelengths approximately 10 μm (micrometers). Infrared also used for non-military operations. Civilians also use infrared in their everyday lives like engaging in short-range wireless communication, environmental monitoring, identifying room temperature and weather forecasting.

Natural Infrared

In an actual temperature of 5, 750 Kelvins, the sunlight is consisted of thermal spectrum, a radiation which is somewhat bigger compared to half infrared. Sunlight, at its pinnacle, delivers an irradiance of 1 kilowatt per square meter at sea level. The energy is composed of 527 watts for infrared radiation, 445 watts for visible light and 32 watts for ultraviolet radiantly. Closely most infrared radiation that came from the sun is somewhat infrared which wavelengths measured shorter than 4 micrometers. Most thermal radiations, on the surface of the Earth which is very far compared to the temperature on the surface of the sun, are composed of infrared in mid-section of infrared region. In totality, objects expel an infrared radiation crossways of wavelengths. However, only a confined region of spectrum is priority because sensors gathered radiation within a specific width of the band. Maximum emitted wavelengths and in reverse comparative to the totality of temperature of an object, that is in line of Law of Displacement of Wien.

International Commission on Illuminating (CIE) endorsed the three bands of infrared radiations' division.

Abbreviation	Wavelength	Frequency
IR-A	700nm- 1, 400 nm (0.7μm- 1.4 μm)	215 THz- 430 THz
IR-B	1, 400 nm- 3,000 nm (1.4 μm- 3μm)	100 THz- 215 THz
IR-C	3, 000 nm- 1 mn (3 μm- 1000 μm)	300 GHz- 100 THz

The specification scheme of ISO 20473

Designation	Abbreviation	Wavelength
Near- Infrared	NIR	0.78- 3 μm
Mid- Infrared	MIR	3- 50 μm
Far- Infrared	FIR	50- 1000 μm

Scheme used by Astronomers in presenting Astronomy Division of Infrared Spectrum

Designation	Abbreviation	Wavelength
Near- Infrared	NIR	(0.7) to 2.5 μm
Mid-Infrared	MIR	2.5 to (25-40) μm
Far- Infrared	FIR	(25-40) to (200-350) μm

These divisions are subjected to change because they are not accurate and it can vary contingent on the publication. Different temperature ranges and diverse environments in space are observed using these three regions.

The adjoining wavelength to the radiation that can be detected by human eye is the near- infrared. Mid and far- infrared are increasingly progressive from the visible spectrum. The beginning of infrared is characterized through different standards at different range between 700 nm and 500 nm, however the limitation between visible and infrared light is still indistinct. Longer wavelengths assisted irrelevant to scenes irradiated by common sources of light. However, IR lasers and IR LED sources which are categorized as intense near- infrared light can be sensed up to approximately 780 nm, and be observed as red light. Wavelengths measured 1, 050 nm, provided by intense light sources, can be perceived as a pale red radiance; resulting difficulty to near IR lighting of scenes in the dark. Leaves when exposed to near- IR appeared to be bright and filter are blocked when all visible escapes from all over the IR, thus eye will be given a second to adapt to the exceptionally blurred image coming through a cloudy IR.

Ultraviolet Waves

Ultraviolet or UV is one of the electromagnetic waves that consists of wavelengths which measurement ranges from 10 nm to 450 nm. The wavelength of UV is smaller compared to visible light but longer when compared to X- rays. Radiation emitted by UV comprises almost 10% of the overall light expelled by the sun, and its presence can be felt from sunlight. Mercury vapor lamps, black lights and tanning lamps also produce ultraviolet, however ultraviolet from the sources cannot be weighed as an ionizing radiation due to lack of energy on its photons to convert atoms into ions. Ultraviolet radiation with long wavelengths can set off chemical reactions and may result to substance to glow or incandesce. Therefore the biotic reactions of UV are grander compared to effects of simple heating. UV radiation's collaborations with organic particles resulted to its many practical applications.

Short History of Ultraviolet waves

"Ultraviolet came from a Latin word ultra which means "beyond" therefore ultraviolet means "beyond violet", violet is one of the colors of white light that possesses the highest frequency. Ultraviolet frequency is greater than violet light. Johann Wilhelm Ritter, a German physicist discovered UV radiation in 1801. He observed that rays, which are not visible, past violet at the end of the visible spectrum. It easily darkened a paper soaked in silver chloride, more easily than of violet light itself. Oxidizing rays is the term used by Ritter to give emphasis to chemical reactions and made him to identify it from heat rays. UV was named "chemical ray" and became famous until 19th century. John William Draper named UV rays as tithonic rays which made the words chemical and heat rays be abolished in favor of new terms ultraviolet and infrared radiation correspondingly. Then in 1878, sterilizing bacteria through the use of short wavelength light was discovered. It was proposed that the range of most effective wavelengths were approximately around 250 nm. In 1960, ultra violet's effect on DNA was recognized. Then it was Victor Schumann, a German physicist discovered the ultraviolet having a range blow 200 nm and called vacuum ultraviolet because air absorbed it strongly and easier.

Electromagnetic Ultraviolet Radiations Spectrum

Name	**Abbreviation**	**Wavelength (nm)**	**Photon energy (eV, aJ)**	**Alternating waves**
Ultraviolet A	UVA	315- 400	3.10- 3.94 0.497- 0.631	Long wave cannot be absorbed by the ozone layer.
Ultraviolet B	UVB	280- 315	3.94- 4.43 0.631- 0.710	Medium wave consumed by the ozone layer.
Ultraviolet C	UVC	100- 280	4.43- 12.4 0.710-1.987	Short wave, destroy germs, consumed by ozone layer and the atmosphere completely.
Near- Ultraviolet	NUV	300- 400	3.10- 4.13 0.497-0.662	Can be seen by birds, insects and fish.
Middle Ultraviolet	MUV	200- 300	4.13- 6.20 0.662- 0.993	
Far- Ultraviolet	FUV	122- 200	6.20- 12.4 0.993- 1.987	
Hydrogen Lymanalpha	# Lyman- a	121- 122	10.16-10.25 1.628- 1.642	Shorter wavelengths, ionized by radiation
Vacuum Ultraviolet	VUV	10- 200	6.20- 124 0.993- 19.867	150- 200 nm wavelength can proliferate through nitrogen, consumed by oxygen in the atmosphere.

Extreme Ultraviolet	EUV	10- 121	12.4- 124 1.99- 19.87	Consumed by the atmosphere completely. Ionizing radiation region.

A different made of solid and vacuum devices have been discovered for the usage of various parts of the Ultraviolet Spectrum. There are attempts to invent light sensing devices, but those devices may malfunction due to unwanted reaction to visible light and different fluxes. Photo diodes and photocathodes can detect ultraviolet which can be fit to be perceptive to various parts of the UV Spectrum. Photomultipliers are sensitive to ultraviolet which are available. The measurement of Ultraviolet radiations are made by Spectrometers and radiometers. Across the spectrum, detectors, which are made of silicon, are used. The lens of the human eye obstruct most radiations brought by wavelength having 300-400 nm. Cornea blocks wavelength that are vulnerable to near-UV and people with aphakia, a condition known to be lacking of lens to observe near- ultraviolet to be whitish blue or whitish violet. Vacuum ultraviolet or VUV has wavelengths which are shorter than 200 nm, are consumed by oxygen, however wavelengths measured approximately 150- 200 nm can proliferate via nitrogen. This spectral range of Ultraviolet can be utilized by scientific instruments through operation in the atmosphere, which oxygen is absent without using the expensive vacuum chambers. Solar astronomy largely drives technology for the invention of Vacuum Ultraviolet instrumentation. Unwanted visible light were removed by Optics because they contaminate the Vacuum Ultraviolet, in broader view detectors' response are confined to non-Vacuum Ultraviolet radiation and solar blind became the most targeted area for research. Vacuum instruments with light cut off photo cathode are appealing when compared to silicon diode. Extreme Ultraviolet is defined by a conversion in Physics of collaboration of matter. Valence electrons of atom interact with longer wavelengths that are approximately 30 nm and the inner shell electrons and nuclei interact mainly with shorter wavelengths.

Ultraviolet and the Sun

UV radiation are emitted by hot objects. The sum emits ultraviolet radiation having various wavelengths, comprising extreme violet where it intersect into X rays at 10 nm. Stars that are vigorously hot emit correspondingly more UV radiation compared to the sun. At the top of Earth's atmosphere, sunlight is consisted of approximately 50% infrared light, 40% visible light and 10%ultraviolet rays, with an intensity ranges about 1,400 W/m^{2}in vacuum.

Sun's Exposure: Its Benefits and Risks

Beneficial Effects

1. Activate Vitamin D

Ultraviolet from the sun helps our body in the production of Vitamin D. Vitamin D is responsible for stronger bones, muscles and boosts immune system. Acquiring colon cancer will be less likely to happen if the body has enough supply of Vitamin D.

2. Treat some skin conditions

The treatment for psoriasis requires ultraviolet radiation. Psoriasis is a condition characterized by peeling off the skin because cells develop itchy and scaly patches. Ultraviolet exposure hindrances the growth of cells and symptoms were calmed.

3. Mood Improvement

Tryptamenes, a particular chemical produced in the pineal gland of the brain, through the stimulation of the sunlight, improves our moods.

4. Aid to animals' vision

Birds, bees and reptiles can see near Ultraviolet light in order to find ripe fruits, flowers and seeds that more attractive compared to the background. The fruits, seeds and flowers appeared differently from the perception of the human eye. Like foe an instance, aided by Ultraviolet rays, some flowers have various markings which may help the bees and birds locate the nectar.

5. Aid insect's journey

Most insects use ultraviolet that is discharged from heavenly objects as references for traveling. The reason why insects are sometimes distracted by light.

6. Disinfection and sterilization

Disinfection and sterilization are the affirmative applications of Ultraviolet. Viruses and bacteria that are present in clothes, nappies, underwear and towels will be killed by ultraviolet once they were hanged and exposed to Ultraviolet radiation. Ultraviolet rays pierce through the cell's membrane, annihilate the DNA and therefore destroyed its ability to reproduce and multiply. It explains why UV rays can be utilized as disinfecting and sterilization of the world,

Risks upon Exposure to UV rays

1. Skin Cancer

UV serves as an environmental factor that can cause cancer. It is the universal agent that can cause skin cancer. It is stated in the studies that the three main types of cancer namely basal cell carcinoma, squamous cell carcinoma are caused by sun exposure.

2. Sun burn

Ultraviolet causes sun burn to the skin. Sun burn is a sign that skin cells are damaged. The skin absorbed too much energy from the sun. In the attempt of the body to repair the damaged cells, extra flow of the blood is doing by the body.

3. Disrupt Immune System

Immune system will be affected by the excessive exposure to the sun. It has an adverse effect on the immune system. Studies show that Ultraviolet can change the allocation and function of the white blood cells which serve as body's protection against diseases caused by virus and bacteria. Exposure to the sun repeatedly may cause the immune system to fail.

4. Damage the Eyes

UV rays can damage the tissues of the eyes, it can cause temporary blindness or snow blindness or photokeratitis, but the blindness will disappear within a couple of days. The journal of the American Medical Association informed that even low amount of sunlight can increase the danger of developing eye damage like cataract, pterygium and pinguecula. Pterygium is a lump on the mucous membrane that conceal the white region of the eye. It is not cancerous but in some cases it protracts to the cornea that may hindrance vision. It does not require treatment but it can be removed if vision may hinder. Pterygium develops mainly on people who live in places with tropical climate and stay longer outdoors. Pollen, smoke, sand and wind are said to be the other causes of the condition. Meanwhile, pinguecula is formed on the eye when the tissue changes and developed a small lump on the conjunctiva of the eye. Pterygium and pinguecula share almost the same characteristics it's just that pterygium has flesh colored form and occurs in shape of oval or elongated, however, pinguecula appears to be yellowish in color and conceived a shape of triangle. Both of them are not cancerous and may be removed if cause irritation and hinder the vision. It is never too late to take care of your eyes.

5. Ages the Skin

Ultraviolet annihilates collagen and connective tissues under the epidermis of the skin, therefore it speeds up aging. Wrinkles, brow line, spots and loss of elasticity are triggered because of the UV rays. The different skin tone of underside and top side of a person signals that he stays longer under the sun. It is never too late to take care of our skin.

6. Fades Color

Ultraviolet absorbs dyes and pigments used to color the objects. Fabrics, paintings and furniture must be protected against UV to prolong its color and life. Take note that fluorescent lamps are also good source of ultraviolet.

X-ray Waves

X-ray consisting of X radiation is also a type of electromagnetic radiation. The wavelengths of most X-rays measured from a range of 0.01 to 10 nanometers, with frequencies ranging from 30 petahertz to 30 exahertz (3×10^{16} Hz to 3×10^{12} Hz). X-rays carried energy ranging from 100 eV to 100 keV. UV rays are longer than those of X-ray wavelengths but wavelengths of X-rays are longer compared to those wavelengths of Gamma rays. Röntgen radiation is also referred to X radiation. Wilhelm Röntgen, a German/Dutch scientist, is credited to be the discoverer of X radiation and gave the name X radiation.

Hard X-rays are composed of high photon energies which is above 5-10 keV below 0.2-0.1 nm of wavelengths while soft x-rays are composed of lower range of energy. X-ray is attributably used to state an image which is produced and that image is called radiographic. X-ray crystallography used to determine crystal structures, this is possible because the hard x-rays have wavelength that are congruent to the size of atoms. However, soft x-rays with attenuation length measured 600 eV, are captivated easily in air.

How can we distinguish X-rays from Gamma rays? Actually there is no agreed definition that can distinguish X-rays from Gamma rays. Based on their source of propagation is one of the ways to distinguish the two electromagnetic waves from each other. X-rays are expelled by electrons while atomic nucleus emits gamma rays. However this definition has several loopholes because other one can propagate photons that are high in energy and the other one is the source of generation is unknown. Analyzing the wavelengths of X-rays and gamma rays is the last option to differentiate them. Gamma radiation expelled shorter wavelengths compared to X-rays. X-rays tubes emit radiation which wavelengths are longer and have lower energy photons. Photons being carried by X-rays have enough energy to convert atoms into ions and unsettle the bonds of molecule. This will result to ionizing radiation that can destroy living tissues. Radiation sickness is related to exposure to very high radiation dose even for a short period of time. Lower dose of radiation can expel a great risk of radiation-induced cancer. In the field of medical aspects, having risk of getting cancer induced by exposing to radiation, the benefit that can get from this type of examination was compensated. Radiation therapy, which uses ionizing capability of X-rays, is utilized to treat cancer by killing malignant cells. Hard X-rays have the ability to cross through objects devoid of being rapt or dispersed. Ocular opaque objects are being photographed to see its inside through the use of X-rays. Radiograph and airport security scanners are most common applications of X-rays in the field of medicine. The penetration depth of X-rays depend on order of magnitude through X-ray spectrum. This regulates the photo energy so application can offer enough transmission traverse the object and clear image of the object. Wavelengths of X-rays are shorter compared to visible light's wavelength, thus probing structure is smaller than the actual view under the microscope is possible. X-ray microscopy used to get high definition images and to define the locations of atoms in crystals, X-ray crystallography is used.

X-rays and Matter

There are three main ways how X-rays interact with matter, these are photo absorption, Compton scattering and Rayleigh scattering. The energy of the X-rays and the primary configuration of the material dictate the power of those interactions. However, the power of each interaction does not depend on chemical properties because photon of X-ray is larger than the energy that binds chemical. In soft X-ray and energy of hard X-rays commend the mechanism of interaction is more on photo absorption or photoelectric absorption. Compton scattering govern in higher frequencies.

Photoelectric Absorption

Photoelectric absorption's probability per unit mass is around comparative to Z^3/ E^3 where Z stands for atomic number and the energy of the instance photon corresponds to letter E. In absorption edges the rule above is not applicable because there are abrupt change in interaction of probability. However, the overall pattern of high absorption coefficient and photon, with low energies with short penetration depths thus atomic number, is very resilient. Photo absorption governs approximately 26 keV photon energy while Compton scattering has taken turns. The limit is greater on greater atomic numbers. The reason why bones are seen clearly on radiographs is that the high amount of calcium and its higher density. The photon of photo absorbed handovers all the energy to the electron then converting the atom to ions which the electron was destined and propagates photoelectron that converts more atoms to ions.

Compton Scattering

The collaboration between X-rays and soft tissues through imaging is called Compton scattering. It is an inflexible scattering of the photon of X-ray made by outer shell electron. Photon's energy is conveyed to the smattering electron, they convert atoms into ions and the wavelengths of the X-ray increased. The photons that are smattered can travel in many location however for the high-energy X-rays the original direction is the same with the direction as it is scattered.

Rayleigh scattering

In Rayleigh scattering the mechanism of the X-ray regime is in the form of flexible scattering. Inelastic or inflexible forward scattering makes the possibility for the refractive index to use, even X-ray is only below 1.

Advantages and Disadvantages of X-ray

Benefits

1. Medical Applications

X-ray can be categorized bone structures. They are used for medical imaging. Projectional radiograph is being used to detect diseases of the skeletal system and other diseases of soft tissues as well. Common examples are; Chest X-ray used to identify lung diseases and Abdominal X-ray is used for the detection of intestinal obstruction. X-ray can also be used for detection of other diseases such as gall stones and kidney stones. Dental radiography is used to identify some oral problems like cavities. Fluoroscopy is a technique for getting an image which is commonly utilized by radiation therapists to gather real-time movement of the internal structures of X-ray source and screen.

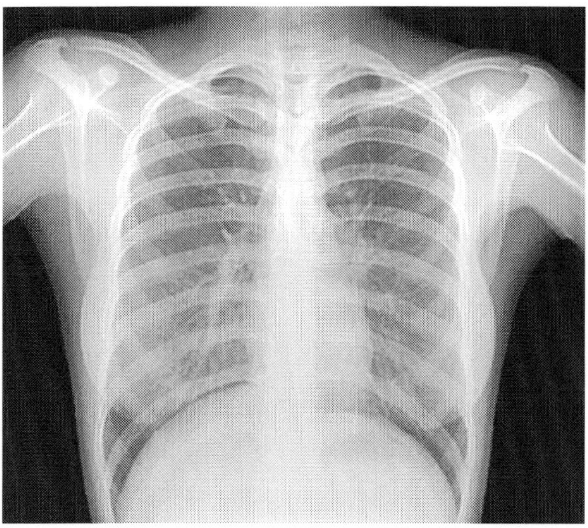

Figure 3.2: An image captured through Chest X-ray.

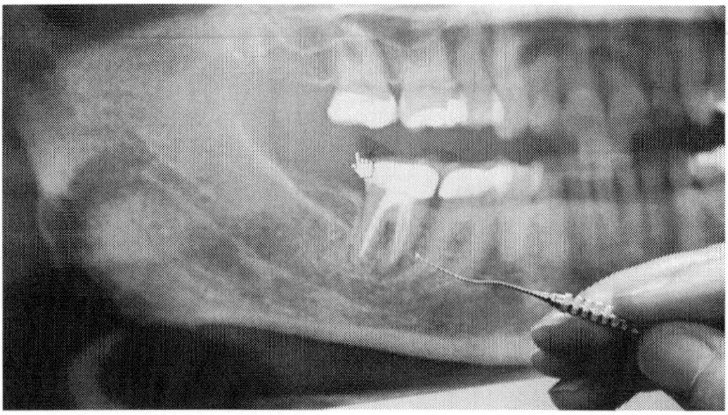

Figure 3.3: Dental radiography showing the oral problems.

Radio Therapy

Radiation therapy is widely used for the treatment of cancer. It needs higher radiation doses compared to those which are used for imaging only. X-ray beams are utilized to treat skin cancer using X-ray with low energy, while X-ray beams with higher energy are utilized to treat cancer inside the body like the brain, lung, prostates and breast.

Harmful Effects

1. Pregnant women must not be subjected to this kind of medical application unless a procedure will be advice by the doctor but it can only be performed after proper assessment if the beneficial factor is higher compared to the risk factor.

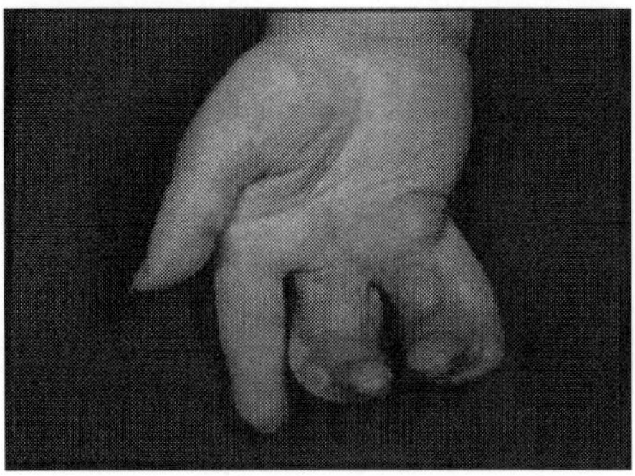

Figure 3.4: An image shows deformity due to X-ray burns.

2. X-ray burns can cause deformity of the parts of the body exposed to X-ray beams.

3. It is theoretically stated that when a person frequently exposed to X-ray radiation there is more likely the increase of risk of getting cancer.

Gamma Rays

Gamma ray also known as gamma radiation, represented by Greek letter gamma (γ), is defined as piercing electromagnetic radiation that arises from atomic nuclei radioactive decay. It is composed of protons with the highest energy range. It was in 1900 when Paul Villard, a French chemist and physicist, exposed gamma radiation. He has studied about radiation that is emitted by radium lead to the discovery of gamma radiation. Then in 1903, it was named as gamma radiation by Ernest Rutherford whom discovered alpha and beta, these two are classified as other types of radioactive decay.

Gamma rays can convert atoms into ions through ionizing radiation therefore it is hazardous to all living organisms. Gamma radiation propagates from gamma decay where in the decay of an atomic nuclei formed from high energy state then decreased to lower energy state. Gamma decay of radio nuclides and radiation from atmosphere contacts with cosmic ray are the natural sources of gamma rays on Earth. There are other earthy sources such as lightning and flashes of earthly gamma ray. Astronomical processes which produce high energy electrons, propagates gamma rays as well, and as a result secondary gamma is produced through bremsstrahlung, inverse Compton scattering and synchrotron radiation. Nonetheless, large gamma rays as filtered by the layers of atmosphere of the Earth and those radiation can be detected by spacecraft alone. Stars as well as the sun are having nuclear fusion where gamma rays are generated but these gamma rays are captivated and scattered inflexible by astronomical objects and have endured their energy before they could escape; that is the reason why they are not observed as gamma rays on Earth. Gamma rays normally have 10 exahertz as frequency range, then having energy over 100 keV and 10 to 11 picometers for wavelengths. Gamma rays no matter the range of its energy, it is an electromagnetic radiation that came from decay of atomic nuclei. The energy that came from radioactive decay has no limit. Gamma radiation consists of several hundred keV and lower than 10 MeV. In the field of astronomy, gamma rays are characterized based on their energy and no particular process of propagation needed from astronomical sources, the energies of gamma rays measured over 10 TeV, which is extremely huge as a result from radioactive decay. Long duration gamma ray explosion is also called as a high energy radiation, it is the energy that can be generated through radioactive decay. Hyper novae is the best example of the bursting gamma rays. Hyper novae is also called as collapsing or exploding star.

Gamma decay is the process where the first gamma ray was discovered. Gamma decay includes a strained nucleus expels a gamma ray quickly as it formed. A French chemist and physicist named Paul Villard uncovered gamma radiation in 1900 at the time he is studying radiation expelled by radium. He differentiated the two radiations resulted from his studies. He noted that the radiation of the unknown source was more powerful than the previous radiation from radium. Radium includes beta rays which is first referred to as radioactivity by Henri Becquerel in 1896, while Rutherford, gamma rays unrepelled by magnetic field, a proof that it is different from alpha and beta. Initially gamma rays suspected to be elements with mass just like alpha and beta rays, that they are just tremendously fast beta, but it was unrepelled by magnetic field indicating that it has no charge. Then it was on 1914, when gamma rays were detected to be repelled from surfaces that are made of crystals, evidently proved that they are electromagnetic radiation. Rutherford together with Edward Andrade calculated the distance of wavelengths of gamma rays based on the wavelengths of radium and they discovered that they are the same to X-rays though gamma rays have shorter wavelengths and higher frequency. Eventually they also accepted that they give more energy per tiny particle called photon. It was understood then that a gamma ray emits a single gamma photon.

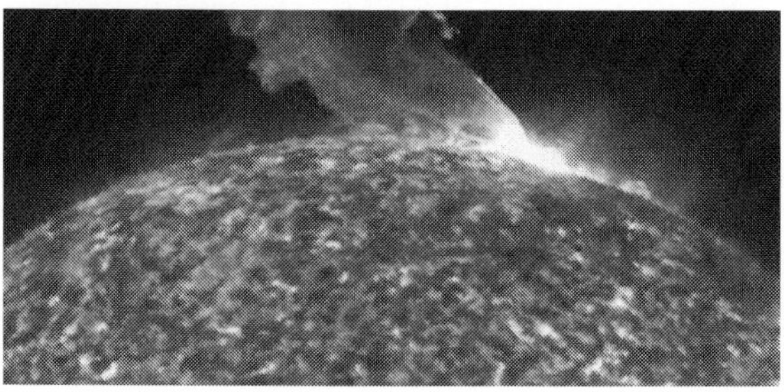

Figure 3.5: Stars including the sun generate gamma rays through nuclear fusion.

Properties of Gamma Rays

Shielding

Shielding that will serve protection from gamma rays involves huge extent of mass, unlike alpha particles can be shielded by paper or skin, and beta which can be blocked by foil. Gamma rays are best absorbed by materials having high density and atomic numbers. Therefore, lead considered as a better shield compare to aluminum, concrete, water or soil because of its compactness due to its adequately higher density. Garments for protection, goggles and respirators can protect from alpha or beta particles but provides no defense from gamma radiation. Therefore the greater the energy of gamma rays, the shielding must be thicker or made from the same shielding material.

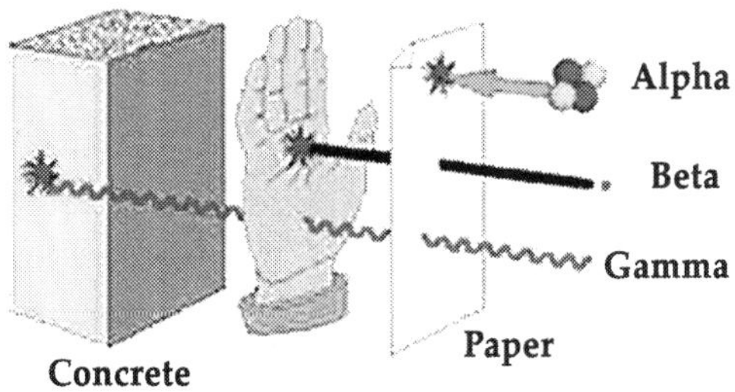

Figure 3.6: The image shows the shielding used for alpha, beta and gamma rays.

CHAPTER 4

RADIATION AND THE EARTH

INTRODUCTION

Earth, from ages ago, has been receiving radiation; it is either solar or nuclear radiation. It can also be either natural or man-made radiation. Life is actually covered with transparent envelop of radioactivity. Radioactivity has occupied a part in our life. There are radioactive materials that are present in the environment in everywhere we go and even on the food that we consume. Air is also composed of some radioactive gases. Human body cannot perform its specific function without these radioactive materials that are naturally present in muscles, bones and tissues.

Radiation is energy that is being conveyed. It is perceived to be in the form of light or particles that are too tiny to be seen. Some sources of radiation emit non-ionizing radiation such as radio waves, visible light, and microwaves, however, X-rays, ultraviolet rays and gamma rays belong to ionizing radiation. Ionizing radiation is considered to be hazardous to all living organisms on Earth. Man was born exposed to radiation either natural or man- made. It could be ascending from the

Earth or from outer space. Cosmic radiation or cosmic rays is the radiation that came from outer space. X-rays is one of the man- made radiation that is used for diagnosing and treating cancer and other diseases that can be seen under radiograph. There are other sources of radiation that man may be exposed to such as nuclear testing, radioactive materials emitted by coal and nuclear power plants.

Though born with radiation on the side, people still feel fear about the hazards that nuclear radiation may cause to human race and to the environment as well. Thirty one countries use nuclear energy to generate electricity but the continuous debate about using nuclear as source of energy still brought controversy to all involved parties. The debate circles mostly about the risks entwined of nuclear against the benefits that can be derived from the applications. Barry Commoner, a biologist and environmentalist, compiled all the arguments against the utilization of nuclear energy. According to him, a nuclear reactor is a monument that people need to care with attentiveness and awareness because after thousands of years that reactor will produce huge bulk of radioactive wastes. Meanwhile, Hans Bathe, physicist and 1967 Nobel Prize winner, is on the other side of the argument. He approves of the process of nuclear power plants as another source of energy. He explains that the only essential source of energy is nuclear fission in place of fossil fuels which require thousands of years to reproduce. Moreover, Bethe insisted that the percentage of hazards that the operation of nuclear power plants are just small compared to other danger that the society consents.

Let us dig deeper about this radiation and find out why people fear it and others perceive hope in it.

Figure 4.1: An early nuclear plant which uses atomic energy to generate electricity (left), a modern nuclear plant (right).

AN ERA WITH RADIOACTIVITY

Radioactivity refers to the particles which are emitted from nuclei as a result of nuclear instability. Radioactivity is defined as the radioactive substance conveyed from unstable atomic nuclei. A radioactive atom is continuously discharge energy or radiation. The unstable nucleus convert other nucleus or change energy level thus radiation was emitted. Just think of a big lump of magnets moving very fast and because of its high, energy some pieces will shoot out and collide to wall or anything. During the operation, as the nuclear energy is applied, materials became unstable and decay will be produced. As the radioactive materials decay they develop energy or waves which can cause damage to cells of all living organism and is hazardous to all life on Earth. Marie Curie, who won Multiple Nobel awards, died because of radiation which he chose to study. The concepts of radiation was not known then, that they do not know radiation can cause serious health problems and hazardous to biological aspects on Earth. In line with her death, her papers and other documents such as her cookbook are sealed in boxes because they are highly radioactive and may impose danger to person who will handle them.

Radium was discovered to be stronger by two million times in radioactivity compared to uranium. Radium salts glow because of radiation. The combination of radium salts and zinc sulfide and glue resulted to glow-in –the -dark paint, this recipe was used by US Radium Corporation. It was used in the hands of wristwatches used by military and other instrument for them to read even in night time. The company hired young women called the Radium Girls, who painted the hands of the wristwatches. In order to paint the tiny hands of the wristwatches, they point the brush using their tongue and lips, not knowing the hazard and danger they put in their bodies. They also decorated their nails and teeth to become attractive in the eyes of their boyfriends. As a result, they swallowed large quantity of radioactive materials, and the reaction began to toil as they experienced anemia, bone fractures, jaw necrosis, fallen teeth and even having still born babies. The women filed a lawsuit against the US radium. Though they won the suit filed against US Radium, they lived for a short time, actually the last girl lived within two years after the suit has settled. US Radium continued its business under safety laws up to 1947.

Radiant Dial Corporation had the same experience with US Radium. They also hired women with the same condition like in US radium and unfortunately the same result happened to the girls. They use radium in almost all the products from cosmetics, toothpaste, cigarettes, suppositories, household cleaning products and even virility booster that when a man put it under his scrotum he will be explosive in bed.

Now because of its discovery, the concepts about radioactivity has revealed. The awareness of its properties has allowed human race to flourish in using radioactive materials that can be beneficial and limiting the risks that may impose to all living organisms.

The studies for materials that discharge radiation has started after Roentgen discovered X-rays in 1895. Wilhelm Conrad Roentgen, a physicist, unable to explain the characteristics and the source of the invisible radiation that he had discovered. Unsure about this radiation, he named it X-ray though. He also uncovered that X-rays can be propagated by pointing a cathode ray upon the walls of a glass tube. Some properties of X-ray were also identified such as X-rays can expose photographic film; that X-rays can move across or penetrate on paper, human flesh and in thin sheets of metal.

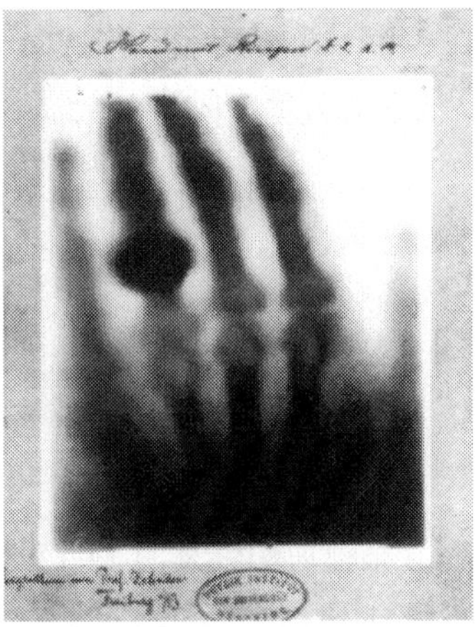

Figure 4.2: The X-ray image of the hand of Roentgen's wife, Anna Bertha Ludwig.

Antoine Henri Becquerel, a French physicist, was discovered the proof for radioactivity for the first time. He was one of the scientists who are determined to look for other invisible radiation. In 1896, he suspected that fluorescent substance may expel X-rays as well. Becquerel used non-phosphorescent potassium uranylsulfate- a uranium salt. Becquerel used a thick black paper to wrap a photographic plate and he put a fluorescent uranium salt above the paper then exposed it to sunlight. He then did confirm that uranylsulfate expelled radiation because of the beclouded photographic film. He kept his work in a drawer because he cannot make some experiment due to cloudy weather but he was amazed when he developed the plate after a few days. He observed the blackened silhouette of the uranium salt. Becquerel did his set up again but instead of exposing it to sunlight, he made it inside the dark room. He discovered that uranium salts expelled radiation even without the sun. Becquerel's discovery treated with disinterestedness because it cannot produce an application until Marie Curie, Becquerel's student in doctoral, studied uranium for her doctoral study. Marie Sklodovska Curie confirmed the observations of Becquerel through several experiments that she had conducted. She observed that Becquerel's rays were still similar under several circumstances such as uranium being dry or wet, solid or pulverized and if it pure or mixed with other elements. She attested that Becquerel's observations were true, that higher quantity of uranium the greater radiation it conveyed. Marie hypothetically guessed that Becquerel's ray was possibly originating from the atoms of uranium. In order for her to test her hypothesis, she decided to search for another radioactive materials.

Then in 1898, Marie exposed that thorium also expels radiation. Thorium's emission of radiation seemed similar to that of uranium, that radiation was discharged from the atoms. She created the word radioactivity, which came from Latin word for ray, in order to define the behavior of thorium and uranium. Pierre Curie, the husband of Marie Curie, joined her wife in her studies. He was fascinated by Marie's discovery that he left his work about crystals. The research conducted by Curies revealed that two uranium ores were expelled greater radiation compared to pure uranium. Because of this discovery they can forecast the occurrence of radioactive elements in the ore. After working under terrible conditions for forty-five months, they finally separate two other radioactive elements. The first radioactive element which is 400 times further radioactive compared to uranium and display properties comparable to bismuth, but due to its radioactive characteristic therefore it was not bismuth. They named it "polonium" came from Poland, where Marie was born. The name "radium" was given to the second radioactive elements that they discovered, radium was determined to be two million times more radioactive compared to uranium.

Pierre and Marie Curie showed the results of their researches and presented the vial of the glowing radium salt in Paris where the first international physics conference was held in 1900. Implications of this new discovery were considered by the scientists. Becquerel, Marie and Curie received the 1903 Nobel Prize in physics.

Alpha, Beta and Gamma

Ernest Rutherford arrived at the Cavendish Laboratory when he was only 24 years old. He came from New Zealand. Inspired by Becquerel's work, he also made a research about Becquerel's rays and discovered two types of radiation emanating from those rays. One type is easily absorb and the other is more piercing radiation. He named the first one as alpha (α) radiation and the other one is beta (β) radiation.

Alpha Radiation

Alpha radiation was composed of dense, positively charged conveyed by atoms of uranium and radium. The occurrence of alpha radiation can be determined when atom experiences radioactive decay, releasing off an alpha particle composing of two protons and two neutrons. Alpha particles interrelate powerfully with matter and take a few millimeters away in air. A thin sheet of paper or even the surface layer of the skin can stop alpha radiation. Yet if a material with a presence of alpha radiation is accidentally taken inside the body through breathing, eating or drinking, wherein internal organs are bare and may cause biological injury. It as Alexander Litvinenko who became a famous example as he ingested and poisoned by polonium-210 in his tea.

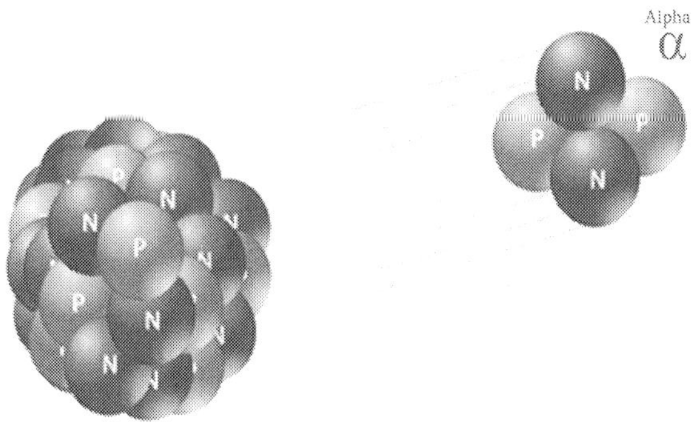

Figure 4.3: Alpha radiation.

Beta Radiation

Beta radiation assumes the form of an electron or a positron (a particle with the size and a mass of electron but charge positively) being conveyed from an atom. They are more piercing compared to alpha particles, it can travel across 1-2 cm of water but an aluminum with few millimeters thick will hindrance the beta radiation from penetrating.

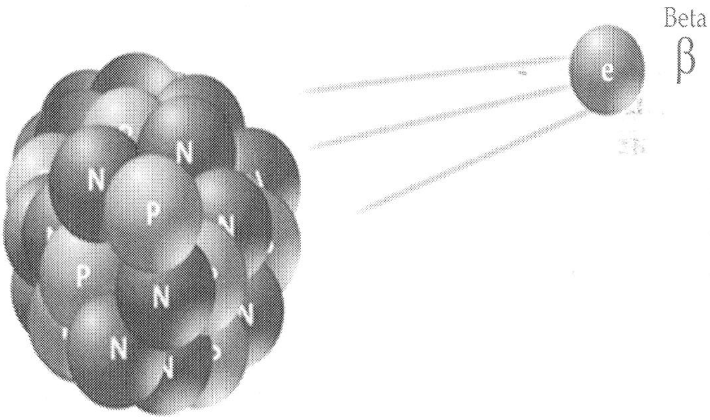

Figure 4.4: Beta radiation emission.

Gamma Rays

Gamma ray also known as gamma radiation, represented by Greek letter gamma (γ), is defined as piercing electromagnetic radiation that arises from atomic nuclei radioactive decay. It is composed of protons with the highest energy range. It was in 1900 when Paul Villard, a French chemist and physicist, exposed gamma radiation. He has studied about radiation that is emitted by radium lead to the discovery of gamma radiation. Then in 1903, it was named as gamma radiation by Ernest Rutherford whom discovered alpha and beta, these two are classified as other types of radioactive decay.

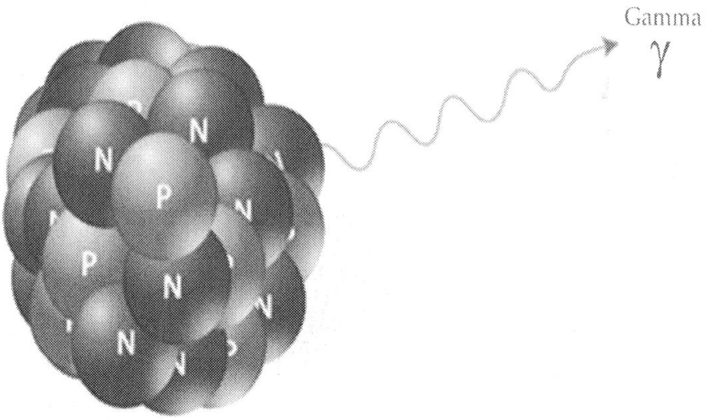

Figure 4.5: Gamma rays.

ISOTOPES: NATURAL AND ARTIFICIAL

The experiment conducted by Rutherford in alpha scattering resulted to the proposal that an atom was composed of nucleus, a positive core. The element's atom is electrically neutral, it means that the nucleus consists of protons and neutrons and electron's number is equal to the number of protons that exist outside the nucleus. Frederic Soddy and Ernest Rutherford suggested that radioactivity may give birth to new elements. This is the reason why other physicist and chemists uncover different elements. However, those elements came out to be the same in chemical composition to known elements but they have different atomic masses. For instance, it was found out that uranium x, ionium, and radiothorium cannot be isolated from ordinary thorium even though they have different atomic masses. These elements conquer the same space in periodic table though they have different atomic masses.

Radioactive Isotopes

Soddy, in 1913, suggested that an element may compose of different forms and consists of various atomic masses. Soddy named the other forms of the given element isotope. After sometime, it was revealed that elements consist of two or more isotopes. Isotopes may be stable or unstable just like hydrogen. The two stable isotopes of hydrogen are common hydrogen and deuterium and tritium is its unstable isotope. Unstable isotopes are radioactive elements, there are different 40 elements that are radioactive. These elements are called radioelements.

Stable isotopes are the nonradioactive elements. It was J.J. Thomson who discovered the proof of multiple isotopes. He conducted streams of neon ions placed in charged and magnetic ground and take down all the measurements of the deflection through a photographic plate. It can be observed in the plate that the stream made a shining blotch. Thomson perceived that on the photographic plate showed two isolated blotches of light and proposed that there are various axis of symmetry of refraction. He therefore resolved that neon gas has atoms that were composed of greater mass compared to the rest.

Nuclide is the term used to name the nucleus of an atom. The symbol below is used to denote nuclides.

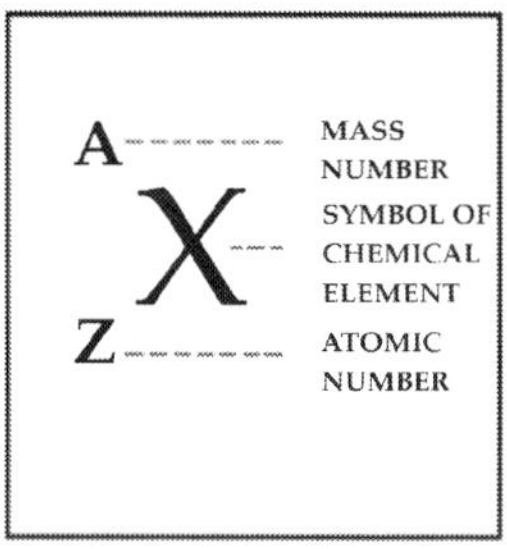

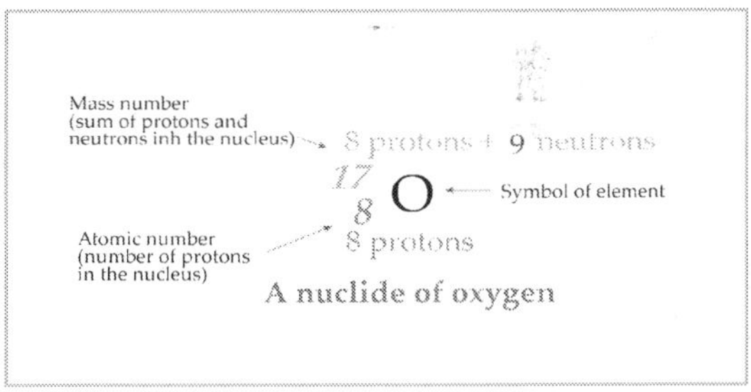

A nuclide of oxygen

The atom of nucleus that has neutron and protons that are even in number are recognized as stable. However, those of elements consisting of even number of neutrons and odd number of protons, protons that are even in number and neutrons in odd number and both having odd numbers are known to be unstable. There are 1,800 different isotopes that were detected, 264 of these are identified stable. Out of 264 isotopes, there are 157 which consist of even number of protons and neutrons. The interaction between protons has strong nuclear force. It is attractive but short in range about 10^{-15} m and at isolation range in zero. Meanwhile, the repulsive electrical force is recognized to be long in range, though it depends on the in reverse of isolation's square of distance. The nuclear force maybe small compared to the electrical force of repulsion, which is in the case of atoms having heavy nuclei. Radioactive isotopes that are naturally occur on Earth, referred to as primordial radio nuclides. Primordial radio nuclides means these isotopes are prevailing since the beginning of Earth. It includes the following examples; tritium, beryllium -7 and 10, magnesium-28, carbon 14 and sodium-22 and 24 and many more.

Radioactive isotopes can also be generated through artificial way, through the utilization of Controlled nuclear reactions. They have been used comprehensively in researches concerned with biological, environmental and medical purposes.

NUCLEAR TRANSFORMATION

The nucleus of an atom is involved in nuclear radiation. The discharge of alpha, beta and gamma rays alters the condition of the nucleus in various ways. The emissions of alpha and beta can change the number of neutrons and electrons in a nucleus and it may lead to production of new element. This unstructured process is known as the radioactive decay. Radioactive decay or nuclear decay is the progression of unstable atom of the nucleus falloff some energy by expelling radiation such an alpha particle, beta particle or gamma rays. Radioactive material contains unstable nuclei. Emissions through proton and neutron caused the decay of particle exceedingly agitated short-lived state of nuclear. Radioactive decay is not a constant process within a single atom. Evidently, quantum theory stated that it is difficult to forecast when a specific atom will shootout an energy unrelatedly how old is the atom. It produces a decrease in mass that are at rest, upon the disintegration of energy and get away in various ways. Decay energy is characterized by difference on the masses of parent nuclide and decay products, this can only be applied to measurements of rest mass where energy was expelled from product system.

FUNDAMENTAL PROCESS INVOLVED IN RADIOACTIVE DECAY

Alpha Decay includes the expulsion of two neutrons and two protons from the radioactive element's nucleus. Just like radium-226, the element that the Curie uncovered, radium continuously experiencing an alpha decay but after expulsing an alpha particle, it cannot be considered as the nucleus of the radium anymore. Based on the periodic table of elements, the element with atomic number 86 reveals that it is radon's isotope with Z= 86 and a mass number (A) of 222 (86+136). The process of decay is shown as this:

$$^{226}_{88}\text{Ra} \longrightarrow {}^{222}_{86}\text{Rn} + {}^{4}_{2}\text{He}$$

$$^{238}_{92}\text{U} \rightarrow {}^{234}_{90}\text{Th} + {}^{4}_{2}\text{He}$$

Alpha Particle

α Particle

Alpha Decay

Beta Decay

Beta decay is a radioactive decay involves the expulsion of beta ray and neutrino from the nucleus. For an instance, the discharge of an electron, the beta decay of neutron change it into proton, or on the contrary through the expulsion of positron, proton is transformed into neutron. Beta particle and neutrino are not present in the nucleus previously before the decay, they are produced on the progression of the decay. Through this, more nonradioactive proportion of protons are acquired by the unstable atoms. Binding energy determined the possibility of nuclide that undergoes beat decay. Nuclear valley of stability is the other term referred to the binding energies included to all known nuclides form.

Other method of creating the nuclei stable is that when one of the neutrons will be changed into a proton by expelling a beta particle containing high speed electron.

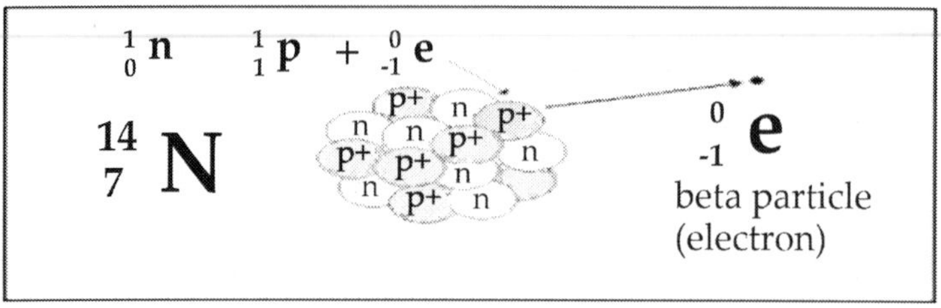

The expulsion of an electron which is characterized by the figure below is one of the processes of beta decay. A stable isotope of helium was a beta decay from Tritium, a meekest form of isotope.

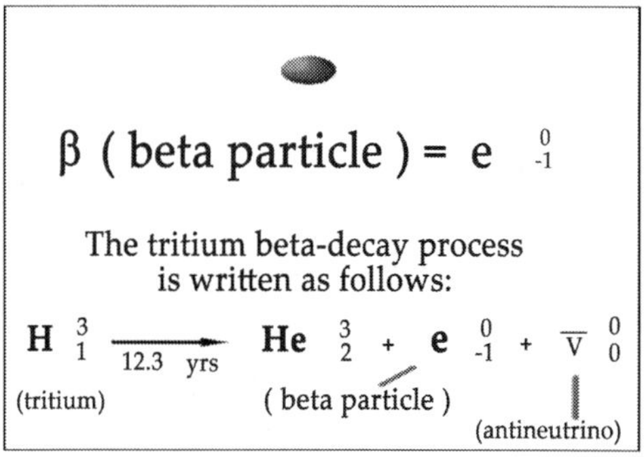

Gamma Decay

In gamma decay, in the contrary to alpha and beta decay, happens when a nucleus is in agitated state and contains extremely large quantity of energy. The energy in the form of gamma rays is expelled as the gamma decay. In gamma decay, only the energy is emitted so the number of protons do not change therefore the decay does not convert the atom into another element.

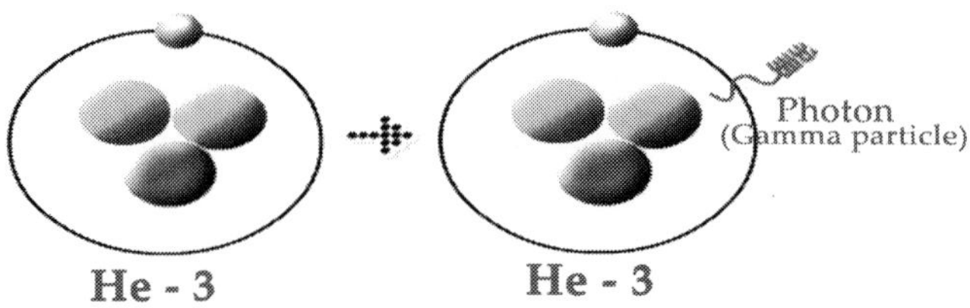

Figure 4.6: Gamma Decay.

DECAY CHAINS

Decay chain defined as the connected sequence of transformations of different radioactive decay outcomes. Radioisotopes do not undergo decay to stable state right away, they submit to sequences of decay until they reach a stable state of isotope. Decay stages are composed of relationships from previous stage. Daughter isotope was formed when a parent isotope decayed. An example is uranium which has atomic number of 92 and decay into thorium with an atomic number of 90, then the daughter isotope may undergo a process of decaying and may produce a daughter of its own; if it may not reach its state of being stable. Granddaughter isotope is the term given to refer to the daughter of the daughter isotope.

A single emission may result to a stable isotope. Other radioactive nuclei submit to a sequence of nuclear transformation before they could get a stable isotope. The best example is the uranium-238 which experienced 15 continuous decay procedures before finally ended to a stable lead-206 isotope. There are four types of decay chain namely uranium series, thorium series, actinium series and neptunium series. They are also known as the heavy disintegration series wherein their products are responsible for the terrestrial radiation's background.

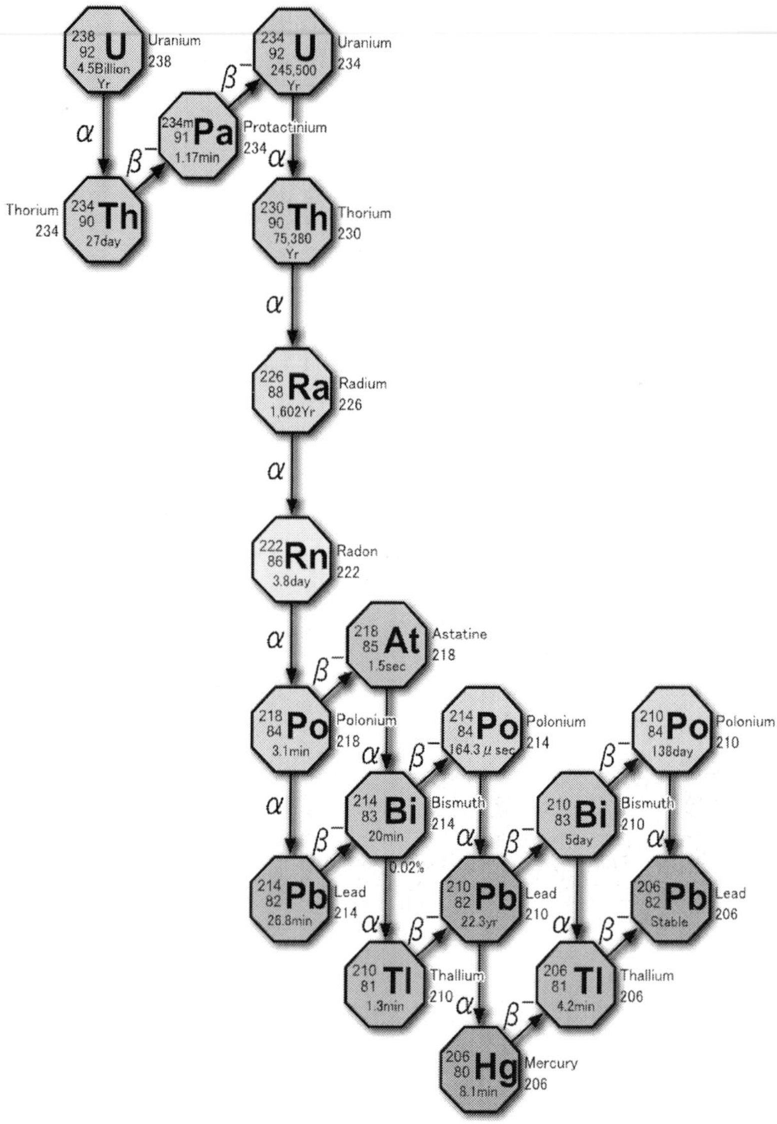

Figure 4.7: The decay chain Uranium series.

HALF-LIFE

Hal-life is the best term to define and explain radioactive decay. The isotope's half-life refers to the interval of time when half of the atoms break down from its original state. This is the term used to define how fast an unstable atom undergoes decaying, how long an atom reaches a stable state after decaying. It is utilized to describe exponential and non-exponential decay.

Table 4.1: Shows quantity reduced as the function of half-lives elapsed.

Number of half-lives elapsed	Function remaining	Percentage remaining
0	1/1	100
1	1/2	50
2	1/4	25
3	1/8	12.5
4	1/16	6.25
5	1/32	3.125
6	1/64	1.563
7	1/128	0.781
……	……	……
n	$\frac{1}{2}^{n}$	$100 (2^{n})$

Formula used to express half-life in exponential decay

Half-life ($t_1/2$)

- An exponential decay process can be described by any of the following three equivalent formulas:

$$N(t) = N_0 \left(\frac{1}{2}\right)^{\frac{t}{t_{\frac{1}{2}}}}$$

$$N(t) = N_0 e^{-\frac{t}{\tau}}$$

$$N(t) = N_0 e^{-\lambda t}$$

- The three parameters are all directly related in the following way:

$$t_{\frac{1}{2}} = \frac{\text{In}(2)}{\lambda} = \tau \, \text{In}(2)$$

Where:

N_0 = initial quantity of substance

N_t= quantity remaining not decayed yet

t ½= half-life of the decaying quantity

Ʈ= lifetime of decaying quantity

λ = decay constant of decaying quantity

Half-life is essential and continuous property of radionuclide. Temperature, pressure or chemical reactions cannot modify the principle of half-life. Nuclear processes can be change by a common physical and chemical changes. Stability of radionuclide is measured by half-life. The smaller the half-life the greater the nucleus decays and the powerful its expulsion. Unstable nucleus is relatively associated with short half-life. The type of decay and half-life will be given consideration in choosing a particular nuclide to be used. Disposal of nuclear waste is as necessary as being knowledgeable of the nuclei's half-life.

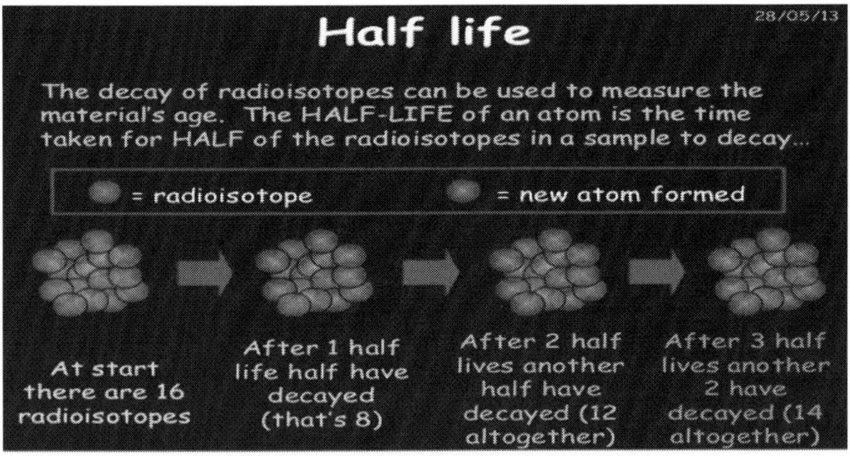

Figure 4.8: It shows the process of Half-life.

UNITS IN MEASURING RADIATION

Human body cannot sense the occurrence of radiation, the utilization of special equipment may tell the amount of radiation that your body has received from source of radiation. Therefore, you do not have the knowledge if you have received a lethal dose of radiation. In order to measure radiation there are different units to be used to define the amount of radiation. For example, you want to measure the strength of any radioactive source given, so the activity is described as the number nuclei that decays or disintegrates per second. The unit that will be utilized is Becquerel (Bq) = 1 disintegration per second. Radiation that ionize will be calculated using volts, erg and volts. The moving electrons are associated with unit named electron volt eV. The isolation of electron from proton requires energy because the electron is strongly confined to hydrogen atom. It requires 13.6 eV of energy to isolate electron away from the proton. Radioactivity in environment and biological are expressed in picocuries per liter and Becquerel per cubic meter.

Radiation doses are measured using rad an acronym for radiation absorbed dose. One rad is equivalent to 100ergs/gram. 100 rad is congruent to 1 Joule/kg, wherein one Gray is also possesses the same value. Roentgen calculated the amount of ionization of radioactive decay that happened in nuclei. One roentgen is similar to 0.93 rad. One roentgen which is equivalent to 0.87 rad that would be the equivalent in air.

Table 4.2: Some Units used in Measurement of Ionizing radiation and Dose

Unit	**Description**	**Equivalent**
Rem (roentgen equivalent man)	A unit of equivalent absorbed dose of radiation which takes into account the relative biological effectiveness of different forms of ionizing radiation, or the varying ways in which they transfer their energy to human tissue. The dose in rem equals the dose in rad multiplied the quantity factor (Q). For beta and gamma radiation, the quality factor is taken one that is rem equals rad. For alpha radiation, the quality factor is taken 20 times rad. Rem is essentially a measure of biological damage. For neutrons, Q is typically taken as 10.	Rem=rad X Q
Sievert (Sv)	A unit of equivalent absorbed dose equal to 100 rem.	1Sv= 100 rem Sv = Gy x Q

Rad (radiation absorbed dose)	A unit of absorbed dose of radiation. Rad is a measure of the amount of energy deposited in tissue.	1 rad = 100 ergs/ gram
Gray (Gy)	A unit of absorbed radiation dose equal to 100 rad. Gray is a measure of deposition of energy in tissue.	1 Gy = 100 rad
Curie (Ci)	The traditional unit of radioactivity of one gram of pure radium-226	1Ci=37 billion dps =37 billion Bq
Becquerels (Bq)	The standard international unit of radioactivity equal to one disintegration per second.	1 Bq= 27 pCi
Disintegrations per second (dps)	The number of subatomic particles or photons released from the nucleus of a given atom over one second. One dps= 60 dpm (disintegrations per minute)	1 dps = 1 Bq

PEOPLE AND RADIATION

All life on Earth experience development and evolution with radiation everywhere. Some countries may prohibit the operation of the nuclear power plant thinking of the hazards and danger it might be brought to human life when radiation leaks from those plants. However, many people live far from nuclear reactor but cannot avoid the exposure to radiation. As mentioned earlier, there are natural and artificial source of radiation. Approximately 87 % of the dosage of radiation is received by the body annually. These 80% mainly originates from natural sources. Four million years ago, physicists perceived that Earth and the living things on it were hit repeatedly by cosmic and terrestrial radiation. The outcome of the cosmic radiation is the interaction of cosmic rays in the uppermost layers of the atmosphere of the Earth. For terrestrial radiation's origin was accounted to the occurrence of radioactive substances that arise from the crust to the upper or surface layer of the Earth. It is said that terrestrial radiation rose from the crust to the surface layer therefore all the crops that are yielded from it and the water that flows through the soil contain radioactive substances. There are also traces of radioactive gases on air that we breathe. Radon is one of the radioactive gases that exists on the air. The term external sources are used to describe the radiations that emanate outside the body.

We are also exposed to radiation within the body. Many of the food that we consume are composed of radioactive elements such as the bananas which contain potassium. Potassium is very essential to power up the body yet it contains radioactive potassium-40. Carbon is also present to the foods that we eat. Radioactive isotope carbon-14 was a decay of carbon.

Nuclear reactions shoot out particles with sufficient energy to isolate electrons the atom. Ionizing radiation produced ion pairs that are extremely radioactive. Alpha, beta, gamma rays and cosmic are the types of ionizing radiation. It does not matter to what ionizing radiation you were exposed to but what matters most is the amount of radiation you have received, that concept is called absorbed dose. Absorbed dose may be measured through Grays (Gy) and Sievert (Sv) which gets the given in Gy and multiplies the effective dose in living tissues. An abdominal X-ray is 0.0014 Gy which is categorized as a light dose and no harm will be perceived from this type of radiation. However, if you were exposed to Chernobyl control room after the bombardment then you will be exposed to 300 Sv per hour. The dose will kill you within 1-2 minutes after exposure.

After the discovery of these radioactive materials, even the researchers and physicists who pioneered in radioactivity did not escape the dreadful effects of radiation. Pierre Curie and Marie Curie also suffered with the effects of radiation on living tissues. Pierre held a vial of radium inside his pocket and later found out that it caused severe damage to his skin. Marie Curie died due to malignant blood disease, it was accountable for her exposure to radiation because of her studies. A number of women in time of US Radium and Radium Dial that suffered several diseases because of ingested radioactive radium. Three hundred live were also lost because of the accidents due to exposure of radiation with high dosage. Ionization within the cells are very harmful because ions are extremely chemically reactive and cause sequences of dreadful reactions that may cause undesirable effect to the cells or worst it may lead to the death of the cells.

The whole body when exposed to four gray is lethal and may lead to death of the person exposed to that amount of radiation. Little bit of dose of radiation may be given to the cancer patients who are undergoing a radiation therapy. However, if you will sum up the entire dosage they received during the therapy is a way greater than the lethal dose but because of the cancer cells that the radiation targeted, only the cancer cells affected by it not the whole organism.

Organ Sensitivity

The organs include in the Table 4.2., are the organs which undergo many cell divisions. They are vulnerable to radiation. They are the first to be affected when the body is exposed to harmful ionizing radiation. Examine and associate the sensitivity of the lungs and the thyroid glands, if these two organs will be exposed to the same amount of radiation, the lungs are four times likely to have cancer than the thyroid gland.

Table 4.2. Sensitivity of organs to radiation.

Tissue/ Organ	Sensitivity
Gonads (ovary and testes)	0.25
Breast	0.15
Red bone marrow	0.12
Lungs	0.12
Thyroid Gland	0.03
Bone surface	0.03
All other organs	0.30

Radiation Applications

After Marie Curie had uncovered the occurrence of other radioactive materials, many companies had competed for the patent rights of the application of this radiation in the field of industry. Some of the applications were radium was painted to the dial of wristwatches of the military, because radium glows in the dark. The purpose is that they will be able to see even in the night. Second, the small amount of radium was added to a liquid called miracle drink. These applications were not practiced nowadays because of what happened to the employees who were painting the radium in the dial of the wristwatches. The medical applications of radium are applied after 20 years Marie Curie had discovered it. About half the century people began to harness power from the nucleus of the atom to generate electricity and made military weapons. The nuclear exploded in Hiroshima and Nagasaki located in Japan brought peace application of the radioactivity.

Agriculture, medicine, forensic science, scientific researches and industry flourish because of the production of artificial isotopes. George de Hevesy was the first to do the practical application of radioactivity. He suspected that the food being served at the dormitory, he was a student that time, are recycled. To prove his suspicion, he put a trace of radioactive material on a specific dish. After a few days, when that particular dish was served to them he checked if the food was radioactive through the use of his gold foil electroscope. His suspicion was correct, the food was radioactive. George de Hevesy was awarded later of the Nobel Prize in 1943 and the Atoms for Peace award in 1959.

There are three major areas in the field of agriculture where ionizing radiation are utilized, the areas namely mutation plant breeding, food irradiation and sterile insect technology. Ionizing radiation came to be a great help to the agriculturist in terms of new genetic lines were produced, plants become more resistant to pests and they are more adaptable to severe climatic conditions. The IR 64 the rice variety which was widely grown in tropic countries was produced by exposing its breeder seeds to three mutagens, these are diepoxybutane (DEB, 0.006%), fast neutrons (FN, 33 Gy), and gamma rays (GR, 250 Gy). Gamma rays can also cause mutations to plant, such as the colorful Rosa Hibiscus that are produced because of exposing to gamma rays. Food irradiation is used to preserved food. Instead of subjecting the export goods into vapor heat treatment, the food are brought to the food irradiation facility where gamma rays will utilize to pass through the body penetrating them to sterilize without altering the taste and the texture of the food mainly fruits to be exported.

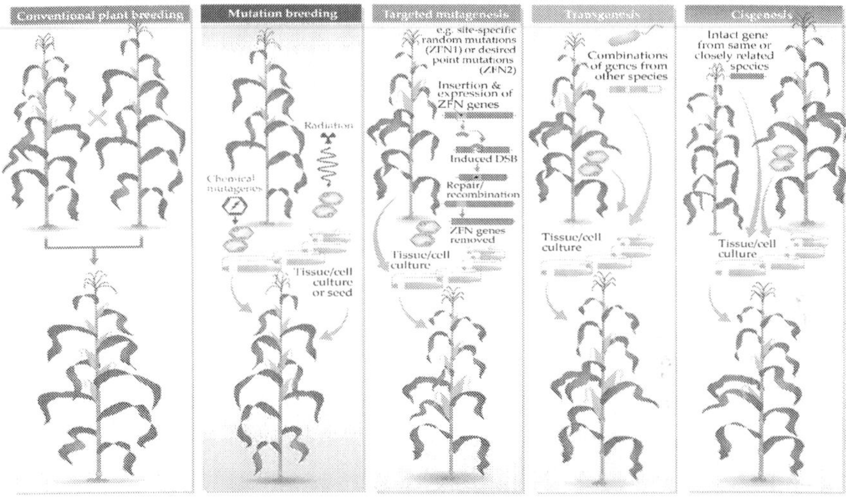

Figure 4.6: Mutation plant breeding.

Food irradiation also helps in the prevention of the growth of the molds in herbs and spices. The use of food irradiation is already growing in almost all parts of the world. Authorities have approved irradiation of several kinds of food from grains products to fruits and vegetables and meat as well. Astronauts when traveling to space consume food that are products or irradiation. The dosage level of irradiation depends on the purpose of preservation, for example if the purpose is to stop the germination of potatoes or onions, a dose of a hundred to a thousand gray will be used to these crops. To eliminate microorganism and pathogens in food it needs a dose of 10, 000 gray is needed. 10 gray is needed in order to make the seeds to germinate. Sterile insect technique is the third application of radiation. The effect of this technique was tested in the island of Guimaras, the application will be used to fight the fruit flies that damage the mango fruits. The scientists irradiate laboratory-raised male insects before hatching. These sterile insects were released in the infested areas and when they mate with female flies, they won't conceived. This technique was used repeatedly and the population of the pests in the area considerably decrease.

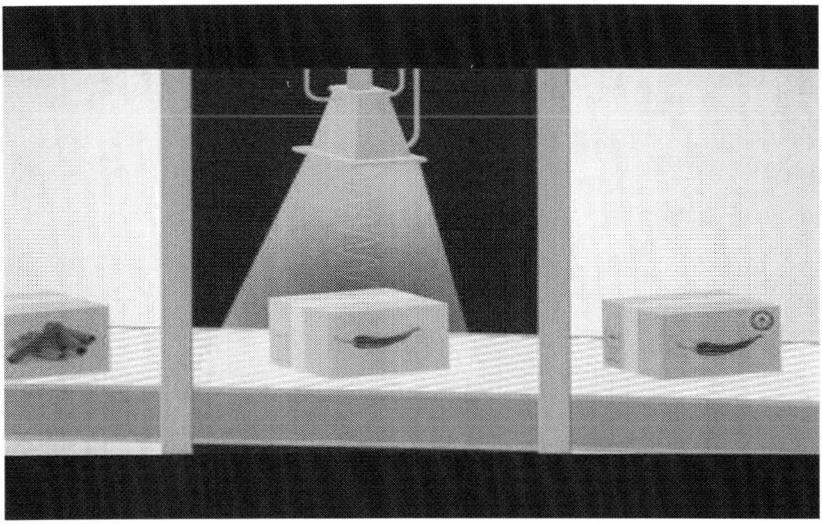

Figure 4.7: Food Irradiation.

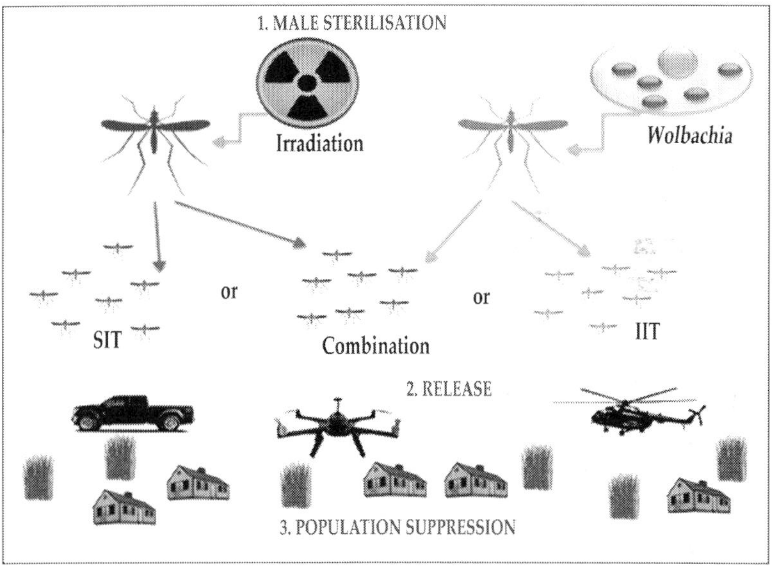

Figure 4.8: Sterile insect technique.

There are many presentations that have been made through the utilization of radioactivity and nuclear energy. Chernobyl and the A-bomb are applications that have created a space in the mind of the people, they won't be forgotten. Some countries are struggling and shelved the option of having electricity generated through nuclear power plant, other countries like France and japan are harnessing the power of atom in generation of electricity. Albert Einstein, a physicists, anticipated the occurrence of atomic age through his theory of relativity. The equation $E=mc^2$ will be used to win the war, that this equation consisted of extremely huge energy locked in a simple matter.

The world's first artificial reactor began its self-sustaining and controllable nuclear reaction in 1942. It was Enrico Fermi and his team assembled and controlled the 0.6-watt reactor. From these inventions, scientists comprehended that nuclear reactors may create energies. That realization gave birth to the first use of fission and fusion of nuclear bombs. Hiroshima and Nagasaki in Japan had tasted the dreadful explosion of the nuclear bomb. After witnessing the effect of nuclear bomb, no one attempted to use the bomb ever again, aside from some nuclear testing of other countries. However, these nuclear weapon testing has an adverse effect in the environment. Nuclear winter are being feared to fallout after the explosion of the atomic bomb, then realization hit military strategists that there are no safe winter in war of nuclear.

Figure 4.9: Nuclear reactor that Fermi and his team made in 1942.

After the controlled fission of U-235 made possible by Fermi, it was comprehended that fission is a feasible substitute to the use of fossil fuels in energy generation. Nuclear reactors was the term given to an equipment that sustain a fission sequence reaction. In the middle of 1950s the engineers created a nuclear reactor that uses unwanted heat that powers the turbines then the electricity was generated. Nuclear reactor has two kinds, one kind is used for power generation and the other is used for research. The difference between the two depends on the output, fuel and the mode of operation. More than 400 research and medical nuclear reactors are operated around the world that save hundreds of thousands of lives every year.

NUCLEAR REACTOR FUEL

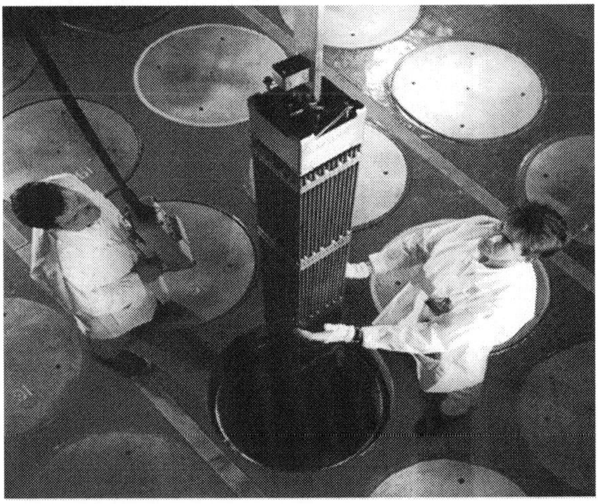

Figure 4.10: Reactor fuel in a reactor rod.

Fissionable materials is required to fuel up a nuclear reactor. The most vital fuels for reactors are Uranium-235 and plutonium-239. Uranium is an example of isotope that is natural and it produces 0.7 % of natural uranium. 99.3 % which remains is non-fissionable and named as uranium-238. Uranium-238 can be transformed into plutonium-239. The commercial grade thermal reactors used uranium fuel that is enhanced by approximately 3% uranium-235 to 97% uranium-238. Fuel rods are used to for the enhanced fuel and be placed in the core of the reactor.

ENVIRONMENTAL HAZARDS FROM NUCLEAR REACTOR

1. Possible accidents

Nuclear accidents may occur when the controls, interlocks and warning systems malfunction. Faulty design of reactor may also lead to nuclear accidents, untrained personnel may be added as a factor of causing nuclear accidents. For example, the Chernobyl accident in 1986 which 56 people died instantly and 4, 000 more are expected to suffer from thyroid cancer and leukemia. Experts stated that in comparison with other accidents and accidents brought by natural calamities, nuclear accident has a small percentage of chance to occur.

2. Routine Radioactive Emissions

The operator of the nuclear power plant had claimed that the reactor has the best design and guaranteed accident-free there are still some radioactivity to the air and the water outside of the plant. ^{95}Kr a fission product which is gaseous state is released to the atmosphere. Water may be radioactive because of the radioactive substances that escaped through water and may result to contamination.

Figure 4.11: Chernobyl accident in 1986.

3. Radioactive wastes

In order to protect the people and environment it is very essential to manage the nuclear wastes. There are three general principles are engaged in the management of nuclear wastes namely delay and decay, dilute and disperse and concentrate and contain.

CHAPTER 5

ELECTRICAL WAVEFORMS AND SIGNAL

INTRODUCTION

From ancient times, electrical occurrences have been the subject of studies, though its theoretical concepts lingered unmoving within seventeenth to eighteenth century. Scientists and engineers had difficulty in designing the application for electricity. The residential and industrial applications of the electricity was established on late nineteenth century. The special characteristic of electricity made it versatile that it has no boundary when it comes to its applications. Applications of electricity include transportation means, heating, communicating, lighting and computation. Electrical power is the framework of industry and society's modernization. The concept of electricity came from the interaction of electric fish with people as they experienced the shock from it. Ancient Egyptians called these fish as Thunderer of the Nile and defined as the guardians of all fish. The nearest and almost discovery of electricity was attributed to the Arabs, they identified lightning and electricity from different sources. Static electricity was studied by Thales of Miletus as the rods of amber brushed to the fur of the cat and he assumed that friction was resulted from amber not from the minerals like magnetite. However, when science developed the bond between magnetism and electricity. Benjamin Franklin was also one of the scientists who showed a deep curiosity about electricity. It was William Gilbert had conducted a study about magnetism and electricity. He created the word *electricus* from the New Latin which means "like amber" and *electron* a Greek word for amber. Luigi Galvani, in 1791, discovered bioelectromagnetics, his discovery showed how electricity mediated neutrons transfer signals to the muscles. Voltaic pile or battery was invented by Alessandro Volta, composed of alternating sheets of zinc and copper. Volta's battery supplied scientists with more consistent source of electrical energy compared to the electrostatic machines that were used previously. Hans Christian Orsted and André Marie Ampere were responsible for the recognition of electromagnetism which deals with the unison of electric and magnetic occurrences. Michael Faraday devised the electric motor in 1821. The electrical circuit was mathematically analyzed by Georg Ohm. James Clark Maxwell linked magnetism, electricity and light.

Late 19th century was the peak of progress of electricity wherein many applications were discovered by famous scientists. Albert Einstein in 1905 published his work about photoelectric effect which explained light energy being transmitted in quantized packets giving energy to electrons. Quantum revolution had begun because of Einstein's work. Photoelectric effect engaged photocells in solar panels that are used to generate electricity commercially.

NATURE OF ELECTRICAL WAVEFORMS

Electrical waveforms are generally ocular illustration of the variation of a voltage or current over time. It means the voltage or current variations be plotted on a graph per which will have x-axis will be the base of time (t) the result will be a waveform. Electrical waveforms are categorized into two distinguishable groups. First, the Uni-directional waveforms where electrical waveforms are constantly positive or negative in its origin, run in one forward path only because these waveforms do not travel in the zero axis reference. Square-wave timing signals, Clock Pulses and Trigger Pulses are the most known Uni-directional waveforms. Second category is the Bi-Direction waveforms which are also known as alternating waveforms because they move from positive to negative direction alternately transmitting constantly at zero axis reference. Bi-directional waveforms experienced periodic changes is amplitude, fairly similar to Sine-wave.

Characteristics of Electrical Waveforms

Any type of waveform whether uni-directional, bi-directional, periodic, non-periodic, symmetrical, non-symmetrical or complex, is composed of three distinguishable characteristics.

1. Period

Period refers to the length of time in seconds when the waveforms assume to repeat itself from start to finish. The term Periodic Time (T) of the sine waves or the Plus Width for square waves.

2. Frequency refers to the series of times the waveform repeats itself within one second time period. Frequency is correlative to time period, (f= 1/T) using Hertz (Hz) as the standard unit for frequency.

3. Amplitude pertains to the intensity or magnitude of waveform signaled measured in volts or amps.

Electrical Periodic Waveforms

Periodic waveforms are the most usual of all electrical waveforms and sine waves is also an example of periodic waveforms. The electricity used in residential is called Sine wave and the application used is alternating current which instantly with a maximum value and a minimum value that alternates over time. Periodic time is the amount of time that a waveform takes between each individual or a round of sinusoidal waveform. Simply the time it takes for the wave form to repeat itself. The period of each waveform depends on the frequency of the waveform. For instance, a sinusoidal waveform takes one second to run its entire cycle will take a periodic time of one second.

Period of the wave pertains to the length of time that waveform requires to run an entire pattern or cycle before it repeats itself. Time period of the wave is measured in seconds so it will be expressed as a period number per second represented by T.

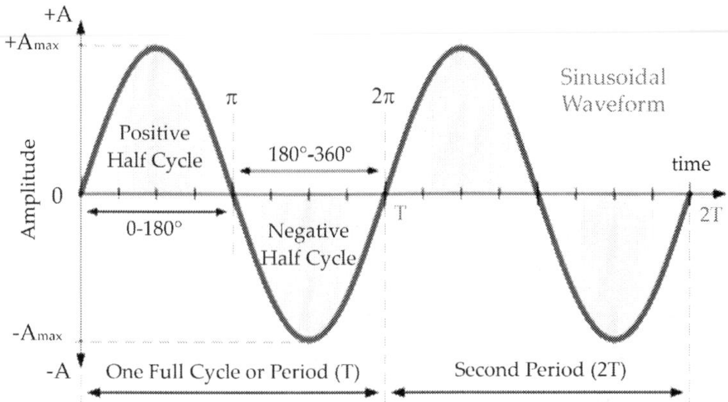

Figure 5.1: The Sinusoidal wave waveform.

Units of periodic time (T) include the following: seconds (s), milliseconds (ms), microsecond (μs). Degrees or radians which expresses the periodic time of waveform, this is in the case of sine wave waveform. One full cycle of radians is congruent to 360° T or in Radians is represented by 2 pi, 2π (T=2π) therefore 2π radians is equal to 360°. Periodic time pertains to the time that electrical waveform take to repeat themselves. Periodic time represents a stable or fixed quantity of time. If the reciprocal of the period (1/T) then the resulted value expresses as the number of times a cycle duplicates itself in one second or the cycles per second known as Frequency. Frequency uses the unit Hertz (Hz) which is characterized as cycles per second (cps) and 1 Hz is congruent to cycle per second. Period and frequency are composed of equation opposite to each other. Therefore as the waveforms' period time decreases, its frequency increases and vice versa.

$$Frequency = \frac{1}{Periodic\,time} or\ f = \frac{1}{T} Hz$$

$$Periodic\,time = \frac{1}{Frequency} or\ T = \frac{1}{f} \sec$$

Figure 5.2: Shows the relationship between frequency and periodic time.

Table 5.1. Prefixes are used to express the sequence of magnitude of the waveform such as kHz, MHz, and even GHz, this is done because 1 Hertz is exactly similar to one cycle per second but 1 Hz is very small unit. (Source: www.wikipedia.com)

Prefixes	**Definition**	**Written as**	**Time Period**
Kilo	Thousand	kHz	1ns
Mega	Million	MHz	1us
Giga	Billion	GHz	1ns
Tera	Trillion	THz	1ps

Electrical waveforms as Square Wave Waveforms

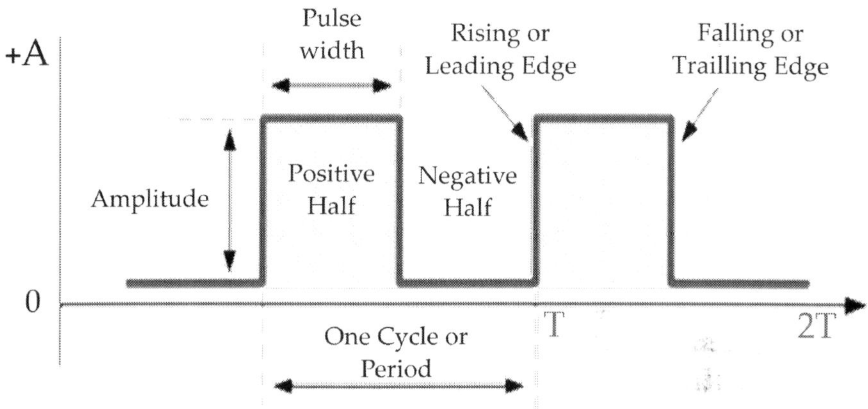

Figure 5.3: Square wave Electrical waveform.

Square wave electrical waveforms are utilized expansively in electronic and microelectronic circuits for clock and timing control signals. They are also known as symmetrical waveforms of congruent and square life span denoting each half of a cycle. Almost all digital logic circuit utilize square wave waveforms on their input and output gates. Sine wave was produced when a wave graphed over time traced by alternating polarity of voltage form a generator. The physics efficacy is the reason why sine wave perceives to be the outputs of electromechanical generator. The proportional rate of magnetic flux and the motion of rotating magnet which voltage had produced because of stationary coils, is changing to the coils perpendicularly. Therefore, when the magnet poles are nearest to the coils the rate is extremely high compared to the magnet poles that are very far from a sine function. The pivoting magnet which follows a sine function, caused changes on the rate of magnetic flux, therefore the coils produced voltage that follows that similar function. Tracing the changing voltage that the coil produced in a generator from a reference on the sine wave graph to that reference when the shape of the wave starts to duplicate itself, then one cycle of the wave will be perceived. The distance between two identical peaks are shown better by spanning the distance between the peaks and the measurements between the resembling points on the graph will be calculated.

The trigonometric sine function of the waveform will be represented by degree marks on the horizontal axis and the angular position of two-pole generator as its shaft revolves. Square waves have bold almost vertical up and down sides with a top which is flat and the bottom produces a waveform that defined similar to its description. Square shaped electrical waveforms are composed of shape that shows symmetry meaning each half of the cycle is identical, therefore the time of positive pulse width and pulse width which is zero, must be equal. Square waves are utilized a clock transmitting in digital circuits with positive pulse width time, this period is called "Duty cycle". The positive or "ON" time of square wave waveform is equal to the negative "OFF" time then the cycle must be 50%.

To sum up, a square wave waveform has shape which is exhibiting symmetry and it possesses a positive pulse width that is congruent to its pulse width which is negative that results to 50% duty cycle. Square wave waveforms are utilized in digital systems to portray a logic level 1 and 2 which corresponds to high amplitude and low amplitude.

Electrical Waveform as Triangular Waveform

Triangular waveforms are typically consist of positive slope which pertains to rise part and negative ramp or decay, making a triangular waveform a 50% duty cycle. The frequency of the waveform will denote the average wave's level of voltage. Lower average voltage level is the result of slow rise and ramp's slow decay time compared to a faster rise and decay time.

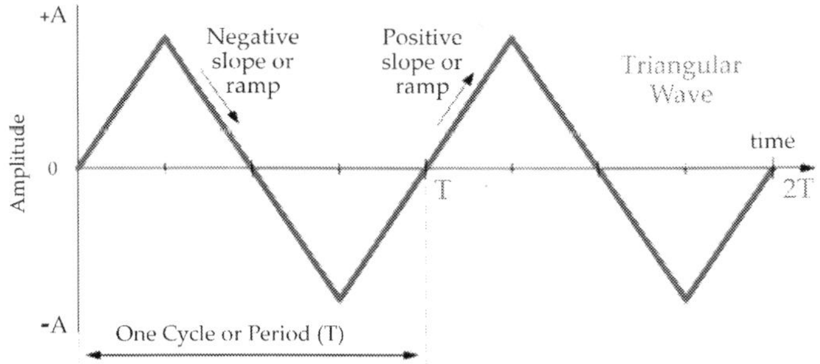

Figure 5.4: Triangular wave electrical waveform.

Electrical Waveform as Rectangular Waveforms

Rectangular waveforms are almost the same with square wave waveforms, the only difference is that the waveform's two pulse width are composed of unequal time period. "Non-symmetrical" is also used to define rectangular waveform.

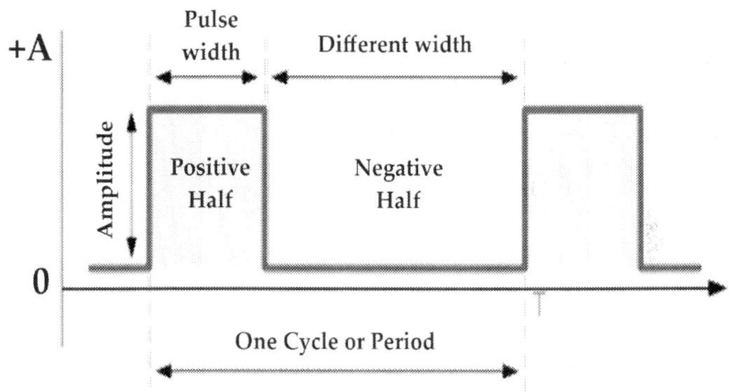

Figure 5.5: Rectangular wave waveform.

Electrical waveform as Saw tooth Waveform

Saw tooth waveforms are another form of periodic waveform. Just like its name, the shape of the waveforms assume the shape of teeth of a saw blade. These waveforms are very similar to each other, they have extremely steep almost vertically steep and slow decay. The more common of the types, this waveform is the positive slope Saw tooth waveform, this ramp almost have perfect linear. The saw tooth waveform consists of frequency (f) and all integer ratio and even harmonics.

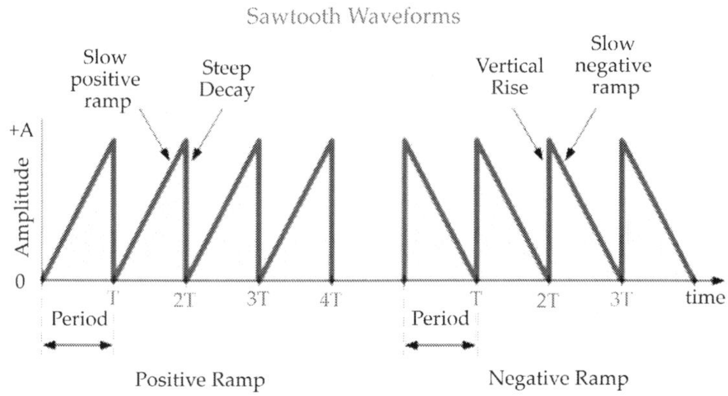

Figure 5.6: Saw tooth Waveforms.

ARBITRARY WAVEFORM GENERATOR

Arbitrary waveform generators is an electronic device utilized to produce electrical waveforms. They can be repetitive or single-shot waveforms which trigger source that required internally or externally. The waveforms which are produced from this device can be inserted into a device that is under testing and analysis of progress through the device being tested, determining the proper function of the device and distinguishing flaws that may appear in it. Arbitrary waveform, unlike function generator, can produce any unparticular defined wave shape as the outcome. It usually characterized as sequence of way points where in particular voltage strike the appearing at specific times in a waveform and the several methods that are inserted which waveform generator can use or leap to those levels. For instance, the duty cycle of square wave in 50% is easily acquired by describing just two points. At t_0, when output voltage is set to 100% and at $t_{50\%}$ set the output voltage back to zero. Just set arbitrary waveform generator to leap between these values and desired square wave will be the result. To compare, set the arbitrary waveform generator to linearly insert between these points and triangle wave will be produced.

The digital signal processing was used by the arbitrary waveform generator to synthesize the waveform, then the highest frequency is always confined to no more than a few gigahertz. Arbitrary waveform generator contains an attenuator, which modulate the waveform result, and has the ability to constantly and repetitively brush the frequency of the resulted waveform between two-operator confinements. This makes the evaluation of frequency response a lot easier in electronic circuit. They also operate like conventional function generators, therefore include standard waveforms like sine, square, ramp and triangle waveforms. Exponential times of rise and fall, sin x/x and cardiac are the units included in built-in waveforms. The users are sometimes permitted to retrieve waveforms from a quantity of digital and mixed-signal oscilloscopes.

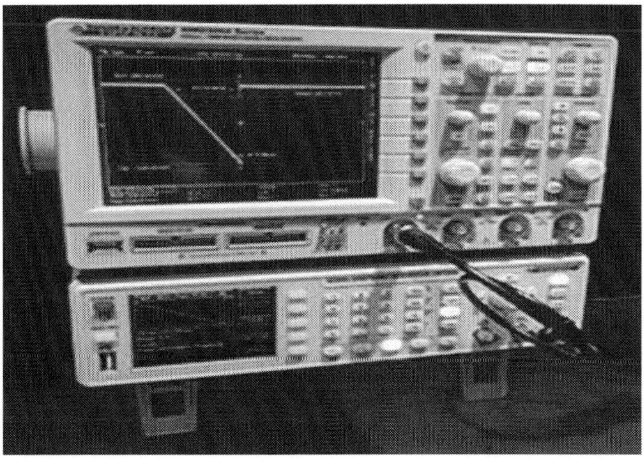

Figure 5.7: HAMEG HMF 2550 digital arbitrary waveform Generator under an oscilloscope showing the produced waveform.

Figure 5.8: The BK Precision model 4078 Dual Channel Arbitrary Waveform

Generator utilizes direct digital process to produce waveforms up to 400, 000 points.

DIFFERENT TYPES OF CHARGE

Electric charge is a result when a force was applied in an electromagnetic field. It refers to the structural characteristics of matter. It has two types of charges these are positive charge and negative charge which are conveyed by protons and electrons respectively. Charges that are alike repel and unlike attract. Neutral is the term used when net charge is not present. An object with excess electrons is perceived to be negatively charge and the opposite pertains to be positively charged or uncharged. Electric charge uses coulomb (C) as its SI unit. Ampere hour (Ah) is used in electrical engineering and in the field of chemistry, elementary charge (e) used as its unit. Charge is represented by the symbol Q which denotes charge. Classical electrodynamics is the other term for how charged substance interact to each other and they do not require quantum effects in solving problem. Subatomic particle has a fundamental preserved property known as Electric charge, which convey their interaction in the electromagnetic field.

Electromagnetic fields are generated and influenced by matter which are electrically charged. Electromagnetic force is the outcome of the interaction between a change in motion and an electromagnetic field. Electromagnetic force is included in the four fundamental forces. Electric charge is proved to be quantized through experiments during the 20th century. When an integer multiples of individual tiny units known to be the elementary charge e. It is approximately congruent to 1.602×10^{-19} C. The proton consists of +e charge and electron is composed of –e charge.

Two Kinds of Electric Charge

1. Positive Charge

A state when an object has deficiency in electrons. It defines as when the area of an object nearest to object that is positive, then there is a net negative charge. Electric current is created when a terminal which is positively charge is attached by a metal wire to negatively charge then electric field is produced by the charges. Positive charge is characterized by more of one type charge than the other on object then the protons determine the positive and the protons in material and the electron with negative charge caused positive charge. The object is said to be electrically charged when the electrons are removed making the object deficient of electrons.

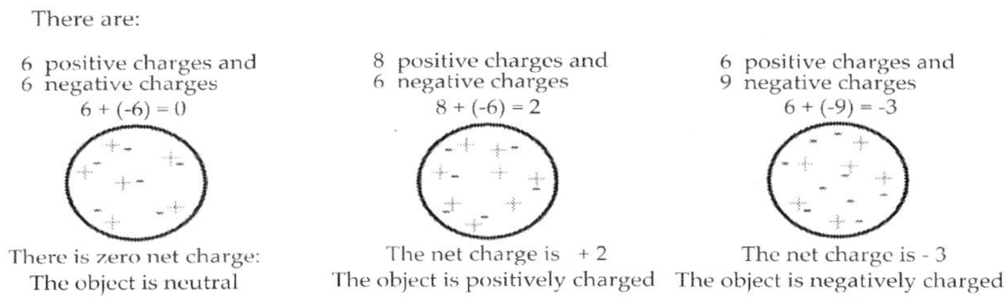

Figure 5.9: Shows the distribution of charges in neutral, positive and negative.

2. Negative charge

Negatively charge objccts are ions that have excessive electrons than their natural elemental form. Electrons are added making the object an excess of electrons.

Neutral cannot be categorized as one of the electric charges because the condition is simply the absence of net charge.

ELECTRIC FORCE AND ELECTROSTATIC LAW (COULOMB LAW)

Electric force refers to the interaction between two charge objects, the interaction may be attractive or repulsive. Newton's law of motion can explain the effect of electric force to the objects. Electric force $F_{elect,}$ is one of the forces that can do something unto objects. It is also denoted as the force that occurs between all charged particles. The occurrences of such phenomena as making the hair strand stand up, chemical bond's creation and be able to see on dark by turning the lamp on are accountable to the characteristics of electric force.

It is said that electric force exists between all particles that have charge. Protons and electrons are extremely small particles that are found inside the atom. Proton is composed of positive charge while electron has negative charge. These are the tiniest charged particles that are present in the environment. All objects that are made up atoms can acquire charges when an imbalance in the number of protons and electrons happened inside the atoms that are exerted in the objects. In the nucleus of each atom, protons are extremely attached to the nucleus itself, so there's no way that they can be moved around. However, electrons are located away from the nucleus and they move freely around within the atom. The movement from one atom to another atom is another characteristic of electron, the transfer of electron from one atom to another causes imbalance to the number of positively charged protons and negatively charged electrons. The imbalance of protons and electrons causes the atom to become charged. When many atoms experience imbalance then the whole object becomes charged.

The best example of interaction between two charged particles is shown when you brush your hair and the hair stand up after brushing it especially on cold and dry days. In brushing the hair, some electrons from the hair are transported to the brush. Because of electrons transfer, the hair becomes positively charged. Now due to two objects hair strands having the same charge, they repel each other causing each strand of hair pushes away from all other, and then hair will stand up. This phenomenon usually happens in a cold and dry day and also possible in hot and humid day. However, if the hair accumulated an ample amount of water, the wet hair just absorb the charges from the air then the charge is likely to disappear quickly. Other similar occurrence is observed when a child slides down on a plastic slide, the strands of her hair become positively charged. The strands repel to each other causing her hair to stand u. the slide had gain electrons because of the brushing of hair to the slide.

Coulombs Law and the Electric force

The amount of charge that each object possesses denotes the strength of electric force between the two charged objects. When the charge's amount gets greater, the force gets greater too. When the distance between two charges become greater the force gets smaller. This process is called Coulombs law and its mathematical equation is shown below.

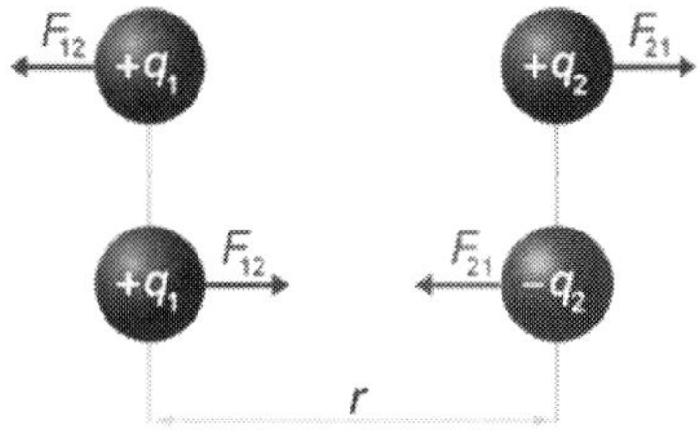

$$F_{12} = F_{21} = k\frac{q_1 q_2}{r^2}$$

Where:

q_1 stands for the amount of charge on object 1

q_2 stands for object 2's amount of charge

kstands for electrostatic constant

F_{12} stands for force of object 1 on object 2

F_{21} represents the force of object 2 on object 1

Voltage

Voltage is an essential quantity that regulates the electricity in the circuit. When electrons travel through a cell or any energy supply, they obtain electrical possible energy. Electrons lose the energy when they pass through the load while flowing around the circuit. The energy transmitted to one coulomb of charge in a battery is known as the electromotive force (emf) of the battery. A cell which is composed of electromotive force of 7 volts transport 7 joules of energy to each coulomb of charge when the charge move within the cell. Potential difference (pd) refers to the energy that is being transmitted to a circuit constituent per unit charge. Voltage is represented by emf and pd. V is used as a unit for voltage in order to give honor to Alessandro Volta.

One volt is the possible variance between points when 1 coulomb of charge between the points was moved by one joule of work done.

1 volt (V) = 1 joule (J)/ coulomb (C)

1 V = 1J/C

In a battery, chemical energy is converted to electrical energy. Electrical energy is also harnessed when mechanical energy was converted by generators. It allows the net flow of unconfined electrons in a circuit. Voltmeter is used in measuring voltage. Voltmeter will be attached to the load which possible inconsistency is to be measured. The positive station of the voltage source will be linked to the positive terminal of the voltmeter and the same with the negative terminal which will be attached to the negative portal of negative voltage source. Unused dry cell labeled 1.5 V can transmit 1.5 joules of energy to every component charge flowing from one station of the cell to the other. The house's electrical lines convey 220 V, therefore sockets at home transmit 220 V.

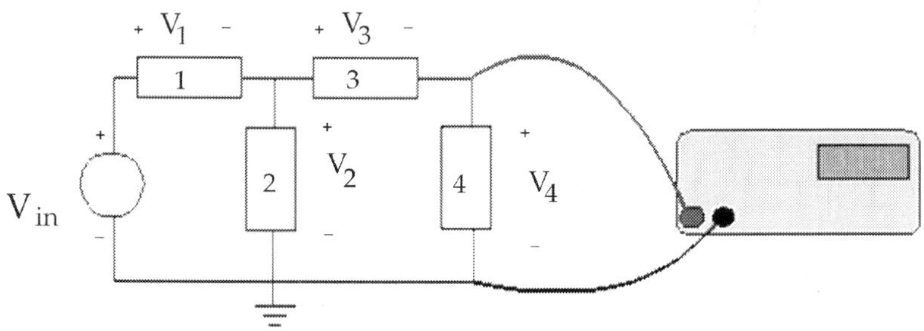

Figure 5.10: a voltmeter attached to two different terminal.

Simple Electric Circuit

A person is accumulating and utilizing the electric circuit whenever he attaches an appliance to the electric socket. Circuit is any procedure of materials that delivers a path for electrons to run. The circuit is composed of electrical source of energy, conducting wire, and a load. The load can be a light bulb, a radio or any appliance. Switch is used in order to turn the circuit on and off.

Figure 5.10: signifies the assembling of electrical circuit in the house.

Electrical energy can be transformed into other forms such as thermal energy and mechanical energy. In the case of the load, it can be transformed into other forms such as light, motion, sound and thermal energy.

ELECTRICAL SYMBOLS

Table 5. 1: Different electrical symbols

Circuit constituent	Symbol
Voltage supply	
DC Source	
1 dry cell	
3 dry cells	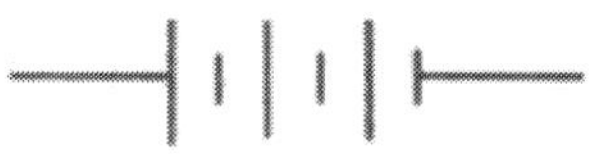
AC Source	
220 V supply	
Example: wall socket	Symbol of AC voltage sourse t

Load Bulb Resistor Other load	  
Switch	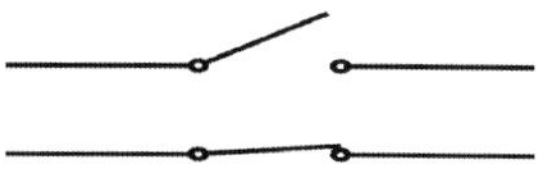
Conducting wire	

ELECTRICAL CONNECTIONS

Electric current is allowed by the electrical connection to flow between points discretely. A pair connection is needed for a circuit. These between points are composed of low voltage difference, the switch controlled the direct current. But, if these points are not linked and the voltage difference those points is high enough. Ionization with regards to electricity will occur and current will occur along the path of pair resistance.

1. Series Connection

In series connection, where there are two loads, the current in the circuit l is as the same as the current through the first load l_1 and is the same as the current through the second load l_2.

$$1 = 1_1 = 1_2$$

Meanwhile, it is also perceived that the voltage across the combination is equal to the sum of voltages cross apart of the combination.

$$V = V_1 + V_2$$

Where V is the voltage across the combination, V_1, the voltage across the first load, and V_2 the voltage across the second load; meaning the voltage supplied is shared by the two loads.

Applying Ohm's law for the first load, $V_1 = l_1 R_1$ where l_1 is the current through the first load, V_1 is the voltage across the load, and R_1 is the resistance of the first load. For the second load;

$$V_2 = l_2 R_2$$

Where

l_2 is the current through the second load

V_2 is the voltage across the second load and

R_2 resistance of the second load

Switching, continuous control and circuit protection and voltage division are the reasons why the series connection is used widely. The main switch inside the series of the home in the electric meter. Then as an equal reaction main switch is constantly in series with the rest of the circuit. The rest of the circuit is always on the side of the fuse.

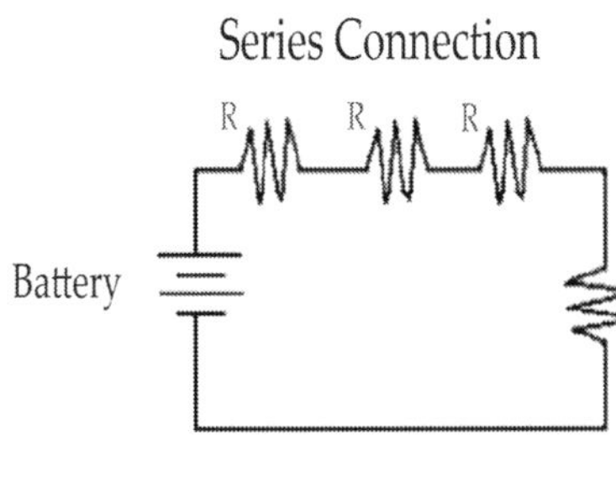

Figure 5. 11: Series connection.

2. Parallel Connection

Parallel connection is associated with two or more loads are linked to a voltage source such a way that the total current is divided among the loads. Current through parallel combination s congruent to the sum of the currents passing through each branch in the combination.

Equation used is

$$I = I_1 + I_2$$

Where I stands for the current through the combination

I_1 the current through load 1

I_2 the current through load 2

Applying Ohm's law to the current combination and individual load

$$I = \frac{V}{R}$$

$$I_1 = \frac{V_1}{R_1}$$

$$I_2 = \frac{V_2}{R_2}$$

These equations are useful in solving for any unknown quantity.

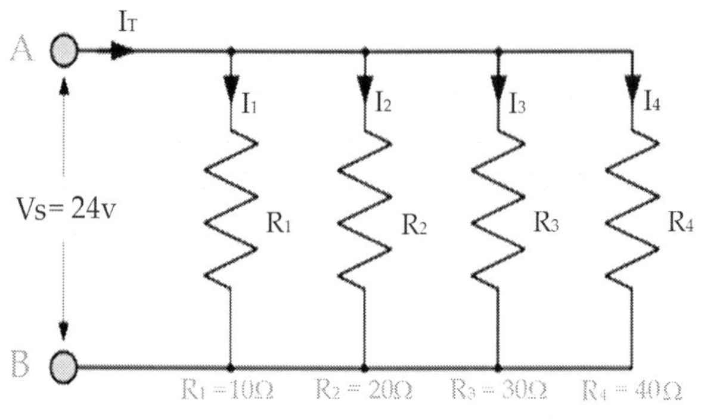

Figure 5.12: Parallel Connection.

3. Series- parallel connection

Occasionally, series and parallel links of resistors are utilize in a single circuit. The end lights at the back of the cars are connected in parallel. Electrical appliances like electric stoves and fans, they have a knob that can make them turn into low, medium and high. The heat of stove and the motion of the electric fan depend where this knob is pointing. These appliances utilize resistors linked in parallel. The resistors in parallel are in turn connected in series with the main resistor of the appliance. Turning the knob meaning switching on or closing part of the circuit.

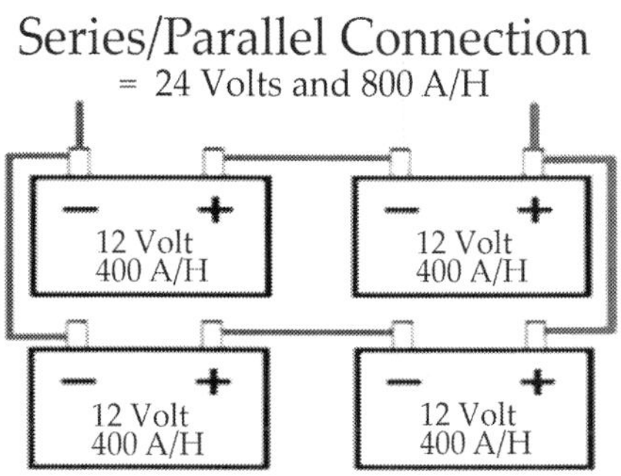

Figure 5.13: Series parallel connections.

CONDUCTORS AND INSULATORS

Conductors

Conductors are objects that allow electrons to flow freely from particle to particle. The object with charge behaves depending on the objects whether it is conductive or nonconductive material. An object that is made of conducting materials will allow charge to be transported across the whole surface of the object. When the charge is transported to the other object at a given location, that charge is quickly distributed to the whole surface of the object. The movement of electron signals the distribution of charge. As the conductors permit electrons to be transferred from particle to another particle, a particle that has a charge will always allocate charge until the entire repulsive forces between excess electron is lessened. If a charged conductor is placed in contact with another object, the conductor can transfer its charge to the object. The transfer of charge in two objects become easily when the second object is also made of conducting material. Conductors permit charge to be transported through the confined movement of the electrons.

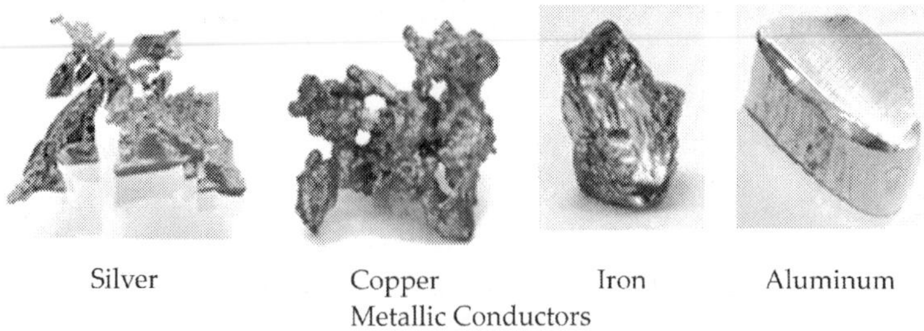

Figure 5.14: Metallic Conductors.

Insulators

Insulators which have the opposite characteristics of conductors, are the materials that hinder the unconfined flow of electrons from atom to another atom and molecule to molecule. If charge is transported to an object which is made of insulator material, the excess charge will stay at the starting location of charging. The insulators particle did not allow the unconfined movement of electrons therefore charge is rarely allocated equally across the insulator's surface.

Though insulators are not good in transporting charge they perform a vital role in electrostatic experiments and demonstration. Insulating objects are used to mount to conductive materials. The insulators being arranged at the bottom part of a conductor to avoid charge from being transported from the conductive object to its surroundings. This also permits the student to operate a conducting object without touching it. The insulators function as handles for moving and touching a conductor.

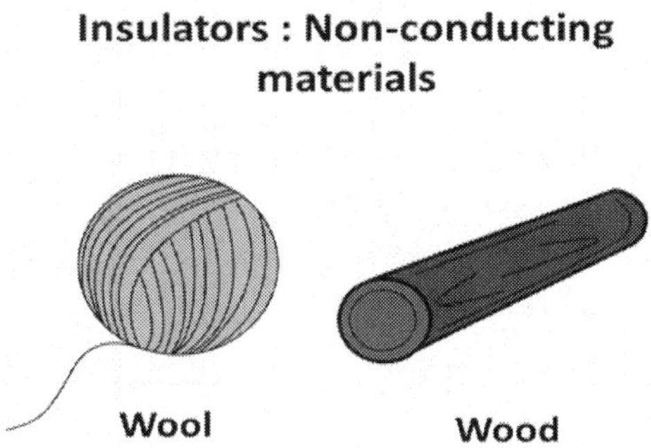

Figure 5.15: Insulators.

Materials as Conductors and Insulators

Different metals, graphite, salts solutions and human body are included in examples of conducting materials. Plastics, Styrofoam, paper, rubber, glass and dry air are the examples of insulating materials. Conductors and insulators are categorized into a division. The materials in this division are arranged in a range or series that are slightly different from each other and that exist between two different possibilities. The materials that are super conductors will be placed at the end of the conductor division and the least conductive will be placed at the other end of the division. Metals are placed near the most conductive and glass will be placed on the opposite end. Metal's conductivity is 3 million trillion times greater compared to glass of insulating division.

In the continuum of the conductors and insulators, the human body can be seen at the middle of the conducting portion. It is perceived that when a body receives a static charge it has the chance to allocate the charge it accumulated throughout the surface of the body. Due to the size of human body, it requires a lot of excess charge before you can see the effect. Van de Graaff generator is used to demonstrate the effects of excess charge on the human body.

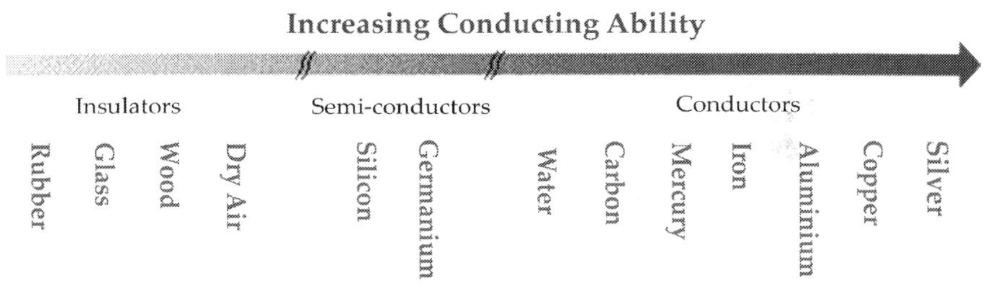

Figure 5.16: Continuum of conductors and insulators.

When someone puts his hand on the static ball, the excess charge is shared from the ball to the human body. The human body being a conductor, the excess charge will be transferred to the body and spread out through the entire surface of the body even on the strands of the hair became charged, and then they repel each other. They put a distance to each other like the neighborhood, the strands of hair starts to rise upward.

Figure 5.17: Van de Graaff generator.

Everybody knows that humidity can do something about high rate static charge. However, static build aps are the most likely to notice in winter months because this is the season with the driest atmosphere. Bad hair day, door knob shocks and static clothing are very likely to happen during winter. During winter months the level of humidity in the air drops to lower quantity. Water is responsible in removing excess charge has the tendency to lose its charges because of the water molecules in the air. The build-up of static charge is more likely to happen during dry air conditions and frequent electric shocks is also experienced. Humidity level varies time to time, day by day and season to season so electrostatic can change from day to day.

ELECTRIC CURRENT AND THE HUMAN BODY

The human body is an example of conductor of electricity. The nerves in its every part are conductors of electricity. Actually, the pulses that transmit along the nerves and generate sensation of pain, temperature, sound, taste and smells are electric in nature. Electric pulses can cause contraction of muscles. Electronic pulses are generated by the brain and other parts of the nervous system in order to function properly, pulses are generated at a regular frequency. These pulses that are being generated inside the body can be detected, recorded and studied in order to obtain records and apply medication if malfunction or diseases may occur. An electrocardiogram (ECG) is a device that record the electric pulses accompanying the beating of the heart. The purpose is to identify illnesses that can be found in heart. It also displays the pulses generated by the brain. If the muscles are strained by the electric shock, the person may be flung away from the circuit. Sometimes the cause of death results from a fall from elevated or high areas. If the muscular contraction causes the victim to be freely moved away from the circuit, even moderately low voltage can be very hazardous. The length of time the body is in the circuit increases the degree of injury. Bear in mind that low voltage does not entail low hazard.

Three primary factors affect the harshness of the shock a person conceives when he/she becomes part of the circuit:

- Quantity of the current through the body (measured in amperes)
- Current's path through the body
- Body's length of time that it became a part of the circuit

Table 5.2: Effects of Current on Human Body

Current level A	**Probable effect on the Human Body**
0.001	Perception level. Slight tingling sensation. Still dangerous under certain conditions.
0.005	Slight shock felt; not painful but disturbing. Average individual can let go. However, strong involuntary reactions to shocks in this range may lead to injuries.
0.006 to 0.03	Painful shock, muscular control is lost. This is called the freezing current or "let go" range.
0.05 to 0.15	Extreme pain, respiratory arrest, severe muscular contractions. Individual cannot let go. Death is possible.

1 to 4.3	Ventricular fibrillation, the rhythmic pumping action of the heart ceases. Muscular contraction and nerve damage occur. Death is most likely to happen.
10	Cardiac arrest, severe burns and probable death.

This Table 5.2 actually gives the information about the general effects of 60 Hz AC electric current on the human body on one second duration.

ELECTRICITY AT HOME: ITS BENEFITS AND HAZARDS

Benefits of Electricity

In ancient times all works of human were accomplished by man himself and the existing machines were limited that can help human through his work load. The progress in science and technology has witnessed this last century, this advancement resulted in the development of numerous machines and techniques that make the human tasks a lot easier. Though these machines need some kind of fuel that keep them in functioning in order for them to help human beings in their work. Ethanol, hydrogen, biodiesel and electricity include the fuel that made the machines to operate. These sources are suitable in their accustomed function, however using electricity as a source for functioning machines and acquiring the comfort of life has its own set of benefits. The benefits that can be acquired through electricity that may bring comfort to human life are:

1. Renewable source of energy

Its availability makes it very useful to mankind. Electricity can be harnessed from renewable sources such as water, wind, and solar. Therefore, electricity is perceived to be renewable source of energy and it does not harm nature in any way. Unlike coal, petroleum and ethanol are nonrenewable resources of energy, aside from they are expensive, once they became depleted it will take millions years for them to be replaced.

2. It does not cause pollution in the environment.

The utilization of fuels like coal, oil and petroleum require combustion in order to apply their function to the machines, in line with this they release a lot of smoke and gases in the atmosphere. The effect of using fuel is hazardous to environment and living beings as well. However, the utilization of electricity as source of energy, it does not give off greenhouse gases in the environment. Electricity that generated from sources like nuclear power plants emit no greenhouse gases.

3. Cost effective

Electricity is more affordable and more cost effective if compared to other sources of fuels like ethanol and methanol.

4. Ease of production

Modern technology in generating electricity are present in almost all the countries of the Earth. There is no country that needs to shoulder the obligations and costs related in purchasing of sources of energy from another country in the world.

5. No need to store

Electricity is produced at power plants and distributed to domestic and industrial utilization. The risk of utilizing electricity has no association in turning on and off the buttons of the appliances that run by electricity.

There are so many advantages of using electricity as a source of energy that it became the most famous and accepted source of energy worldwide.

Electrical Danger and Safety

The dos and don'ts for electrical circuit inside the house.

1. Don't overlook wavering lights, blown fuse or sparks, and light switches and outlets that are hot to touch. Let the qualified electrician check the wiring.
2. Do not overload or what we called "octopus wiring. Read the labels of the electric device and avoid using lamps or bulbs that are not matched with the source of income.
3. Old extensions must not be used. They are used for short term only. If you observed cracks do not use it anymore.
4. Switches and outlets that don't work must be reported to someone who is licensed electrician.

5. Do not use electrical appliance at time of storm and are using cellphones prohibit during the storm. Lat the storm passes.
6. Keep away the kids on exposed wirings. Use safety covers or special outlets.
7. Let the wiring system of the house to be inspected and expect inspection at the right time.
8. Utilize surge protectors if needed.
9. Keep electrical appliances away from the source of the water. Do not touch the water inside the plugged-in heater.
10. Make sure to screw bulbs securely. Loose bulbs may overheat and burn drapes and other light materials inside the house.
11. Do not tug the electric wire when unplugging the appliances, it may cause damage to the wire and may cause electric shock.
12. Unplug appliances when not in use.
13. Do not replace a blown fuse with coin or any wire because it requires correct size of the fuse.
14. When using an electric appliance make sure that you are standing on a wet ground. Electric shock may occur.
15. Do not wear jewelry when working around any electrical appliances.

CHAPTER 6

EARTHQUAKE WAVES

INTRODUCTION

Earthquake waves are seismic waves, these are waves that carry energy and travel through the layers of the Earth. Seismic waves are propagated through earthquakes, volcanic eruptions, magma movement and landslides. Ambient vibrations refer to the low amplitude waves that are created by other natural or anthropogenic sources. Geophysicist who are specifically called as seismologists study seismic waves. Accelerometer or seismometer hydrophone are used to record seismic wave's fields. Density and elasticity of the medium identify the velocity of the propagation of the waves. On the Earth's crust velocity has the tendency to increase in depth and ranges from about 2 to 8 km/s up to 13 km/s in the mantle.

There are distinct types of waves that consist of different velocities that created by earthquakes, various travel times of the seismic upon reaching the observatories assist the scientists to identify the origin of the hypocenter. Seismic waves refraction and reflection in the field of geophysics are utilized for studies and research of the structure of the

Earth's interior and in order to explore, shallow and subsurface structures man-made oscillation are often produced. The discovery of the earth's core and the maps that show the layers of the Earth's inside. Seismic waves can be bent just as the light refracts at its faces then seismic waves bend, reflect and change velocity at limitations area between things below the surface of the Earth. The types of seismic waves are propagated by earthquakes. These types of seismic waves include P (Primary) wave, S (secondary) waves and surface waves which travel and reach the seismic recording stations just after another. Primary wave and secondary wave pierced through the interior of the Earth, however surface waves cannot penetrate the Earth's interior layers. Because of this, body waves include P and S waves. Surface waves reach the observatory last and seismictomographers show least interest because surface waves do not pierce deep inside the Earth, which is the reason why information and details about them are limited.

Many years ago, seismologists observed that information from an earthquake changed when the event happened at distant places, around 105° within the angle between the quake and seismograph at the Earth's center. Because of that distance, waves vanish completely until the slow surface had reached the station. Shadow zone was the term used in defining the area beyond 105°. P waves would arrive with the absence of S waves. Because of this, the scientists came to realize that Earth's core is fluid and molten. The absence of S waves in the shadow zone aid the P waves binding to form this shadow zone.

HISTORY OF EARTHQUAKE WAVES

In ancient times, ancient people believed that giant snakes, turtles, catfish or spiders that live under the ground create earthquakes as they move. They have many nonsensical explanations for earthquakes, they always involve huge and agitated creatures that live under the Earth.

It was Aristotle who attempted to propose explanation regarding the earthquakes based on environmental and natural occurrences. He hypothesized that winds whipping the earth caused vibration of the Earth's surface. Experimental observation of the earthquake's effect were very seldom to occur. People show no interest in studying earthquake. But England was undeniably rocked by a sequence of five intense earthquakes. A cataclysmic shock and tsunami were followed after the earthquake on November 1, 1755 which killed an approximately 70, 000 people. This event knocked out the city of Lisbon Portugal because of the disastrous event most of its people evacuated in churches. The phenomenon had marked the kickoff of the modern age of seismology, following numerous studies researches into the disastrous effects, locations and timing of earthquakes.

Marking the Lisbon earthquake, sages had researched almost exclusively to studies of Aristotle, Pliny and other scientists who conducted studies about earthquake. But as through the years progress the studies about earthquake has developed as well and give emphasis on ideas that are based on modern observations and explanations. Recording the times and locations of earthquakes and observing the physical effects of earthquakes became a top priority and many people led the earnest studies of earthquake like John Michell of England and Elie Berfrand in Switzerland. Increasing studies of earthquake phenomena were heated because of the Lisbon earthquake. The researches aid efforts were often caused by earthquake catastrophe, like an earthquake in 1783 Calabrian which killed 35, 000 people in the southern part of Italy.

The development in the field of communication has been a great help to make earthquake observations throughout the world has been combined. Maria Graham, an author, reported systematic changes in the height of the coastline of Chile, which was after the disastrous earthquake that hit Chile in 1822. Studies in changes of coastline were determined as a follow up observation in the 1835 Chilean earthquake. The confirmation was conducted by the captain of the H.M.S. Beagle named Robert Fitzroy while Charles Darwin was examining the geology of the Andes on shore.

Three Europeans simultaneously made efforts in seismology. An engineer born in Dublin who designed bridges in London named Robert mallet. He calculated the velocity of seismic waves in earth through explosions caused by gun powder. He aimed to find for variations in seismic velocity that would convey different properties of Earth in association with earthquake. His method is still used today especially in oil field exploration. The depth of an underground earthquake was first estimated by Robert Mallet as well. At the time when mallet was plotting gunpowder for explosions in order to study and observe earthquake in

England, Alexis Perrey of France was designing numerous analyses of earthquakes' catalogs. He was researching for periodic differences of earthquakes in association with seasons and with the aid of lunar phases. Luigi Palmieri of Italy devised the electromagnetic seismograph, which was mounted over Mount Vesuvius and the other one is at the University of Naples. Palmieri's seismographs were the foremost seismic instruments that has the capability to detect earthquakes that cannot be intercept by human beings. The invention of these seismographs has set the stage at the end of 1800s and 1900s, when progress in seismology would be harnessed. Three English Professors in japan namely, John Mine, James Ewing and Thomas Gray, who works at Imperial College of Tokyo that time, devised the first seismic equipment that are sensitive enough to be utilized in scientific researches and studies of earthquakes.

Figure 6.1: The early portable seismometer invented by Luigi Palmieri in 1876.

Grove Karl Gilbert of United States concluded that the faults were a primary feature of earthquakes. He stated that after studying the fault scrap from the 1872 Owens Valley, California earthquake. People perceived that earthquakes are the outcome of explosions under the ground and faults were the outcome of the explosion. Harry Fielding Reid of United States modified the work of Gilbert into a step higher. He observed the trace fault of 1906 San Francisco earthquake, Reid concluded that earthquakes are actually the outcome off the partial build-up of stresses inside the Earth's core. These stresses occurred and accumulated after many years, this stress is due to distant forces and is finally released violently during an earthquake. This build-up of strain after many years can cause Earth to rebound.

Japanese researchers also observed the scientific inquiry about earthquakes and they began to conduct studies about it. Seikei Sekiya has become the first person named as professor in seismology, he also analyzed seismic quantitatively. Fusakichi Omori, another famous researcher, studied the decay of aftershock actively following large earthquakes. Omori's equations are still used nowadays. An increased interest into scientific study of earthquakes has begun in the 20th century. The studies and researches about earthquake has become broad because of the contributions made by countries which are experiencing disastrous earthquake. These countries include United States, Japan, Europe, Canada, Russia, Mexico, Central and South America, New Zealand, China, Australia and other countries.

DIFFERENT TYPES OF EARTHQUAKE WAVES

Earth vibrations and destruction are the outcome of three basic types of elastic waves. Only one of the three cannot penetrate to the inner core of the Earth and cannot generate within a body rock. The Primary or P wave is faster compared to secondary or S waves. The motion of the P wave is the same as that of sound waves in that, as it spreads out, it simultaneously compresses and dilates the rock. P waves can travel via either solid rock, like Granite Mountains and liquid material such as volcanic magma or the salty water of the oceans. P waves are also known as the push and pull waves. They are also referred as longitudinal waves. These waves can be perceived like sound waves since both of them are compression-dilation or compression refraction waves. Every particle of P waves vibrate to and fro to the path of wave generation. P waves can travel through gases, liquids and solids in similar way. These waves transport outward from the point of disturbance in all directions in straight lines. P waves are the fastest of all earthquake waves. The average velocity is 5.3 km a second and a maximum of 10.6 km per second. P waves are the first to reach the epicenter and they followed concave path in travelling the earth.

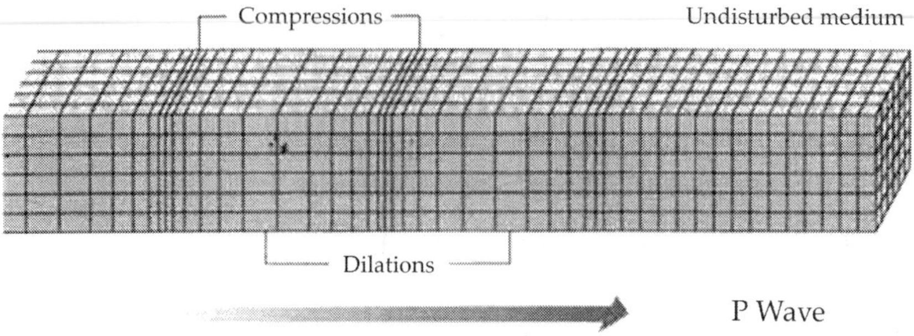

Figure 6.2: The Primary wave or P wave.

The slower wave compared to P wave that travel through the body rock is called Secondary or S wave. S waves generate, its shears the rock sideways at right angle to the direction of the travel. S waves cannot generate in the liquid part of the Earth, like oceans and lakes because if a liquid is cropped sideward or twisted it will not spring back.

Secondary waves are called transverse wave as well. The particles of the waves shake at the right angles to the direction in which they transported with regards to the direction of the propagation. They travel only through solids. S waves cannot travel through liquids. The speeds of travel of P and S waves are different even in the same kind of rock because their speed depends on the variations of properties. Density and compressibility of the rock governed the velocity of P waves, however S waves depend on its density and rigidity. Actually, P waves transport at about 1.7 times the speed of shear waves. P waves are always followed by S waves. Even though S wave's velocity is less than P wave, the S wave is more damaging. Both P waves and S waves cause the shaking of the Earth. The definite speed of P and seismic waves rest on the density and elastic characteristics of the rocks and soils through which they travel. Of all types of earthquake, P wave is perceived first. The result is the same to a sonic boom that bumps and rattles windows. After a few seconds, the S waves arrive with up and down and side to side medium. Vibrating the ground surface vertically and horizontally. This is the wave motion that can cause extremely damages to structures.

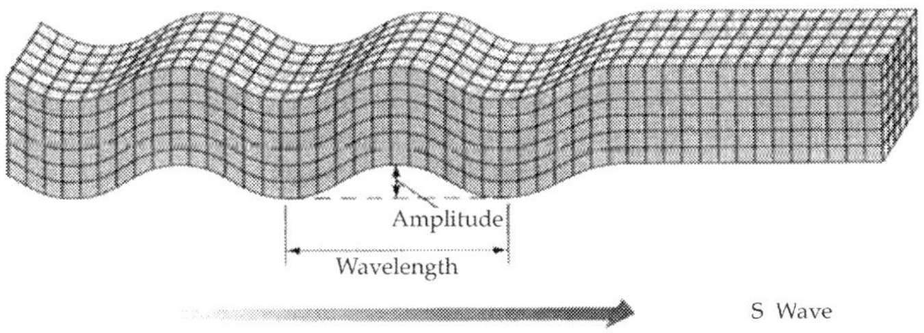

Figure 6.3: The Secondary waves or S waves.

Surface wave is the third general type of earthquake wave. The reason behind is that its motion prohibit to go near the ground surface. This type of waves corresponds to ripples of water that travel across a lake. Surface waves reach the earth's surface after P and S waves. Surface wave's velocity is slower compared to P and S waves. Surface waves also create damaging effect. They provide essential information for differentiating between the continental and oceanic types of crust. Surface wave, which is defined as its motion being limited near the ground surface, is the third general type of earthquake wave. It resembles the ripples of water that transport across a body of water.

There are two types of Surface waves. The first type is called Love wave. Love waves are basically that of S waves that vertical displacement is not present. It transfers the ground from side to side in a horizontal plane out at right angle to the direction of generation of waves. The horizontal vibration of love waves is specifically destructive to the foundations of structures. Rayleigh wave is the second type of surface wave.

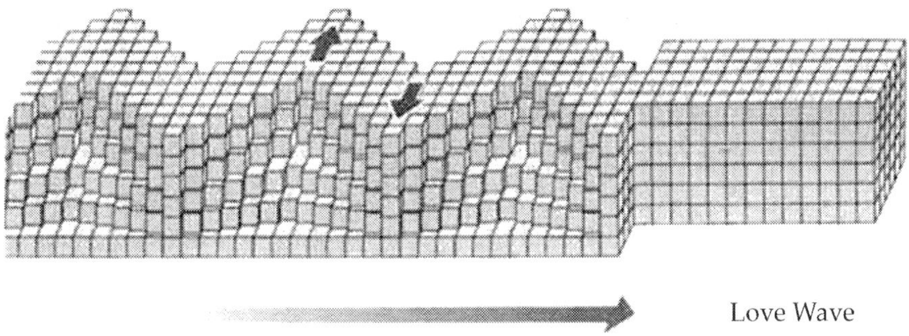

Figure 6.4: The Love wave.

Rayleigh waves move unhurriedly than body waves (P wave and S wave) between the two surface waves love commonly transports faster than Rayleigh waves. Love waves which cannot generate through water has an effect to the surface water only when the sides of lakes and ocean bays push water side ways like the sides of vibrating tank, Rayleigh waves can affect the bodies of water because if its vertical component of their motion.

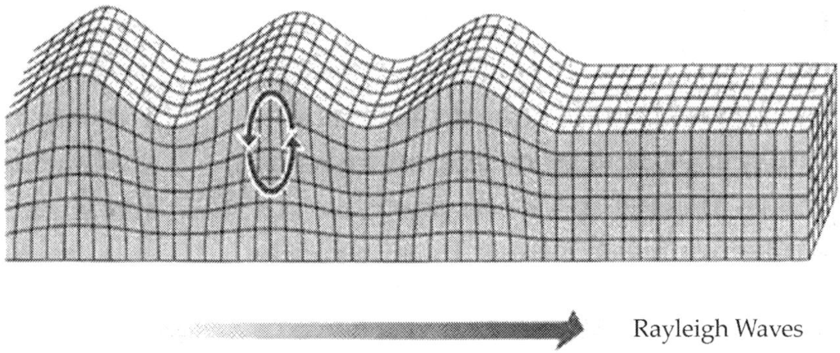

Figure 6.5: Rayleigh wave.

Difference between P waves and S waves

A characteristic of P and S waves that affect vibration, these effects are they reflected and refracted at the interfaces when they travel through the layers of rock in the crust. When either of the waves reflected, energy of the specific type will be transformed to waves of another type. For instance, P wave transports upwards and attack the bottom part of alluvium, a portion of its energy will transfer upward through the alluvium as a P wave and a portion will transfer upward as the transformed S wave motion. Then the portion of the energy will be replicated back downward as P waves and S waves.

EARTHQUAKE OBSERVATIONS

Vital Role of the Investigation/ Observation of Earthquakes

Motivation that came from few branches of science has been a great help in the investigation of earthquakes. It is the time when construction of very sensitive seismic equipment had begun the modern investigation of earthquakes because of this it was possible to record all strong shaking of the Earth's crust. The progress and development of the seismic instruments had made people trace or track the earth with enough information about the intensity of the elastic waves that can be happening after an earthquake. The distribution of information over the earth's surface has not been concede to one uniform plan because the number of stations with seismic instruments are still limited. However, the chance to be given with this instrument still depend on the condition which were fulfilled in one less appropriate place than on other more suitable one. International Seismological Association, which was founded in 1903, has suggested that all civilized nations are engaging to create a methodically organized earthquake stations and to build observations that are based on certain principle.

The recordings that seismic instruments have given up until now have already set an essential light on the nature and properties of earthquakes and the most important is on the nature of movement stages which seismic disturbances are collected, this direction where earthquake waves usually came from and the speed of seismic waves are transported throughout the earth. However the primary value of the modern observation of earthquakes through the use of seismic instruments provide the world a better understanding of the condition of the earth's interior which is completely inhibited from undeviating observations. Investigations using instruments alone is not yet enough to answer most central problems of earthquake investigation. The effort to record all seismic disturbances and publish annual stipulations which tackled all known earthquake disturbances in the form of journal or chronicle was aimed commonly to establish the points or districts of the earth which are prone to shockwaves or tremors and from which earthquake waves discharge over more or less great surfaces.

Due to the insufficiency and incompleteness of the reports of earthquakes in later years, it was difficult to analyze the first seismological problem. It is hoped that within the modern world where communication is possible to all nations, in working hand by hand provide the awareness about earthquake through this, there will be solutions to questions which are formerly impossible to solved.

The Most Significant Earthquake Phenomena

Perceived or not perceived by the human senses, earthquake is the name given to all those tremors. Earthquakes originated from any disturbance of equilibrium within the earth's mass and which are transmitted. They can be transmitted in the shape of spherical waves, as elastic vibrations which includes waves of compression and rarefaction, through the earth's body from their point of origin. Submarine earthquake or seaquake is the term used to pertain to earthquake waves that leave the Earth's crust that lies at the bottom of the sea, and their spherical waves are transmitted across the body of water to the level of the sea.

a. The tremors of an earthquake

The sporadically interchanging compression and rarefaction of the crust of the Earth in asking which are caused by the vibrations of equilibrium in the Earth's interior. The waves of compression and rarefaction of the movement are divided into a vertical and horizontal component. Due to extreme energy strain and movement which is confined in the elastic waves, earthquakes appeared to be sudden tremors of different intensity. The vertical component dominates directly above the subterranean earthquake seat and also in epicenter's neighborhood. The tremors of an earthquake was perceived by the human as a shock below in an upward direction.

b. Number and duration of the tremors

In most cases, the earthquake is compared of a single shock and lasts only for a fraction of second, the outcome will be the most destructive depending on the intensity of the earthquake waves. Whole sequences of tremors which variations in force follow one another at shorter or longer intervals.

Normally, frail tremors occur first then the major tremor come and the end of the tremor is consisted of vibrations slowly becoming weaker and weaker. The seismic vibration can occur with the most resilient tremor then fades with weaker shaking. In this case the overview of the shocks is labeled as the earthquake and the length of the earthquake comprises the time, inclusive of the intervals, which intervenes between its first appearance and the last disturbance. The length of an earthquake is commonly overrated because the observers are surprised by the sudden appearance of the earthquake and typically stay under the imprint of the excitement for a period after its end before they occur to their senses and are capable of realizing of what has happened.

Swarm of earthquakes are terms used to describe the number of shocks which simultaneously occur over a short period is very large. The interval of time over which shocks range may include several days or even weeks and months. Regular earthquake district is the district which is repeatedly visited by earthquakes.

c. After-shocks

A large number of aftershocks will be felt when a very violent earthquake has occurred. The resilient the major shock and the smaller the area that is shaken by an earthquake the more after-shocks. The period over which the after-shocks may include several years. However, when the duration of earthquake is longer, the frequency of the after-shocks included interval that will be lessened. Sometimes the after-shocks that the district experienced does not comparatively correspond to the major shock; the epicenter of the after-shocks transport themselves much more within the major shaken areas.

The people are not concerned with the after-shocks because they give no importance to the aftershocks in contrast to the major shock or tremor. So in time with this people must be informed and aware that they must give importance to after-shocks as well because after-shocks could be the most violent. Therefore, every after-shock must, with respect to time, duration and intensity, be denoted with the same precautions as the first shock. The observation of after-shocks is more essential compared to any other seismic phenomena. The appearance of the after-shocks probably depends on the variations of the air pressure on the area shaken by an earthquake and on the attraction of moon and star. After-shocks best competent to light on the queries of the occurrence of earthquakes.

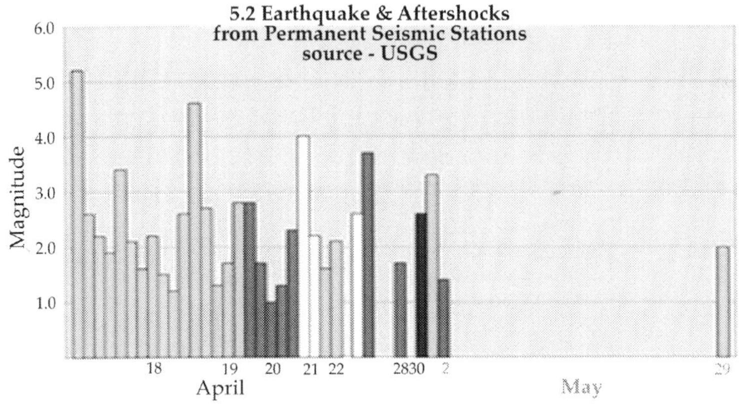

Figure 6.6: shows 5.2 earthquake and aftershocks from permanent seismic stations.

d. Intensity of earthquake

Conventional scale gives a usual force of shocks. The best known and most used is the earthquake intensity which is devised by De Rossi and Forel. It differentiates 10 degrees;

1st Degree

Micro seismic motion, only recorded by seismic instruments

2nd Degree

Shock registered by seismographs, established by a small number of observers who are in state or rest.

3rd Degree

Shock was observed by several persons in a state of rest; strong enough for duration or direction to be estimated.

4th Degree

Shock observed by persons in activity; moveable objects shake like windows and doors, cracking of floors.

5th Degree

Shock is commonly remarked by the whole population; shaking of larger objects, furniture, beds, isolated ringing of house bells.

6th Degree

Commonly awakening of those who are asleep, general ringing of house bells, and vacillation of trees. Isolated cases of persons leaving their houses in terror.

7th Degree

Overturning of moveable objects, loosening of plastic pieces on the ceiling and walls, ringing of church bells, general term but no damaged building.

8th Degree

Falling of chimneys, cracks in the walls of houses.

9th Degree

Part or entire demolition of certain building.

10th Degree

Great catastrophic, run's, fissures in earth's crust and land ships.

Generally one may comment that earthquakes are stronger in the surface than in depths. However, the effect of an earthquake depends in a high degree or the nature of earth's crust involved. It is evidently perceived that one same shock will be very differently under otherwise like conditions in places which are near to one another.

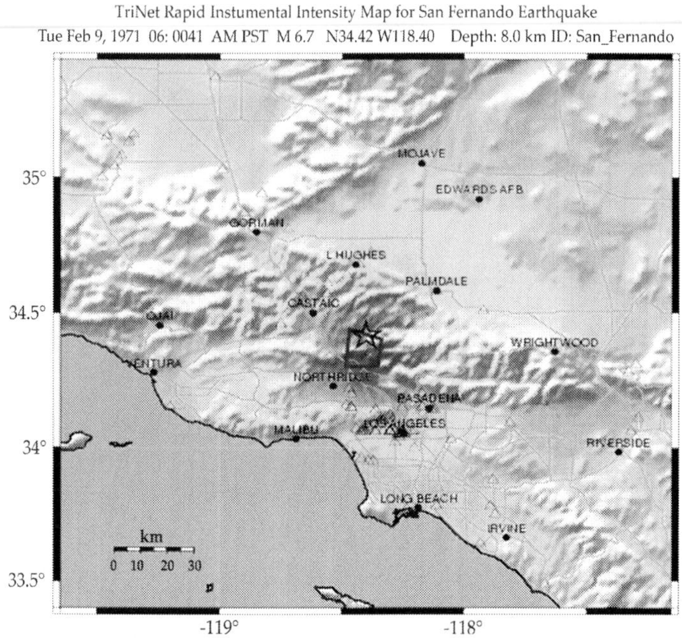

PERCEIVED SHAKING	No 11eH	Weak	Light	Moderate	Strong	Very strong	Servere	Violent	Extreme
POTENIAL DAMAGE	none	none	none	Very light	Light	Moderate	Moderate/Heavy	Heavy	Very heavy
PEAK ACC (%g)	<.17	.17-1.4	1.4-3.9	3.9-9.2	9.2-18	18-34	34-65	65-124	>124
PEAK VEL (cm/s)	<0.1	0.1-1.1	1.1-3.4	3.4-8.1	8.1-16	16-31	31-60	60-116	>116
INSTRUMENTAL INTENSITY	I	II-III	IV	V	VI	VII	VIII	IX	X-

Figure 6.7: Intensity map for the January 17, 1995 Kobe, Japan earthquake.

Effects of Earthquake on the Earth's Surface

Faults, cracks, fissures which run off in the most manifold directions, intersect and thus cut up the land into blocks changes the earth's surface. Normally the occurring formation are round holes which looks like an inverted cone and which may throw forth slimy water when violent earthquake takes place. In this case sand cones which have the appearance of crater are formed. Earthquake may result to movement of masses such as land slips, mountain slips and subsidence, especially the soil is composed of loose water-sodden material. The destruction of buildings is a specific importance for the approximation of the direction of transmission and of intensity of the earthquake waves. The solidity of the buildings in association with earthquake depends predominantly on the materials used in the building, considering its construction.

Major forms of destruction observed in buildings are as follows:

1. Complete or nearly complete ruin of the buildings.
2. Falling in of the gable-walls with preservation of the side walls and of the super-structure of the roof.
3. Preservation of the gable-walls with a partial falling-in of the side walls with the super structure of the roof.
4. Destruction of certain corners, generally the upper ones, and whole ledger of the buildings.
5. Falling –in of the whole wall- work together with sinking of the super structure of the roof.

Figure 6.7: Effects of earthquake.

TriNet Rapid Instumental Intensity Map for San Fernando Earthquake
Tue Feb 9, 1971 06: 0041 AM PST M 6.7 N34.42 W118.40 Depth: 8.0 km ID: San_Fernando

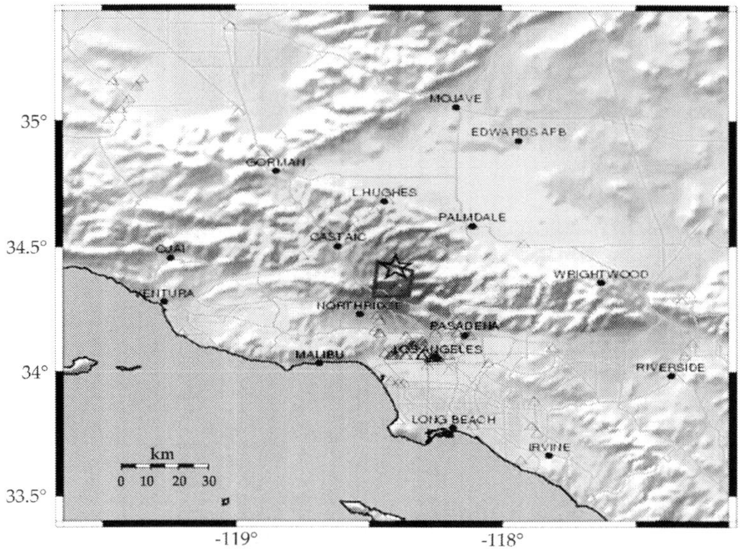

PERCEIVED SHAKING	No 11eII	Weak	Light	Moderate	Strong	Very strong	Servere	Violent	Extreme
POTENIAL DAMAGE	none	none	none	Very light	Light	Moderate	Moderate/Heavy	Heavy	Very heavy
PEAK AOC (%g)	<.17	.17-1.4	1.4-3.9	3.9-9.2	9.2-18	18-34	34-65	65-124	>124
PEAK VEL (cm/s)	<0.1	0.1-1.1	1.1-3.4	3.4-8.1	8.1-16	16-31	31-60	60-116	>116
INSTRUMENTAL INTENSITY	I	II-III	IV	V	VI	VII	VIII	IX	X+

Figure 6.7: Intensity map for the January 17, 1995 Kobe, Japan earthquake.

4. Supplementary Phenomena of Earthquakes

The sound phenomenon is the most important among the phenomena which among other follow in the direction of earthquakes. Earthquake sounds frequently take simultaneously with it and still continue after the end of the quake. Roaring, whistling, howling, rolling, thunder cracking and bellowing are variously said to be the nature of earthquakes. Earthquake sounds occurs in the both earth-and- sea quakes. The force of the noise stands in no relation to the force of shock; feeble shocks may be accompanied by a very loud disturbances and vice versa. Noises are heard in other places without any conveying by a shock of the ground being felt. There are so called ground claps. It has different names in other countries.

PROCEDURES IN REDUCING HAZARDS MADE BY EARTHQUAKE WAVES

Substantial studies and work have been done in seismology to describe the nature and characteristics of the recorded ground movements due to earthquakes. Information about earthquake is required to forecast ground motions in future earthquakes so that structures that are earthquake-resistant can be designed. Earthquakes cause death and destruction through its secondary effects such as landslides, tsunamis, fires, and fault rupture. The violent shaking of the land may result to collapsing man-made structures may associate to losses such as lives and properties. Therefore, the best effective way to lessen the damage brought about by earthquakes from an engineering standpoint is to plan and construct structures capable of enduring violent ground movement.

SEISMIC HAZARD MAP

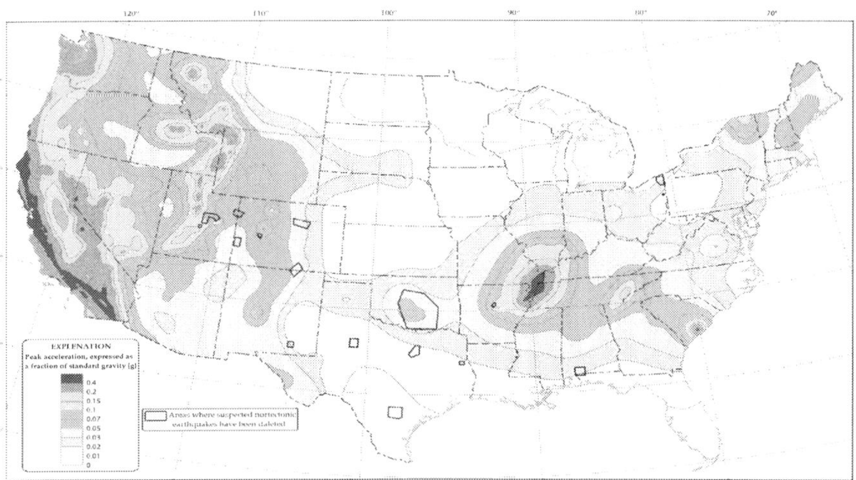

Figure 6.8: Shows Seismic hazard map.

Table 6.1: Foremost Historic Earthquakes

Date	Time‡	Place	Latitude	Longitude	Fatalities	Magnitude	Comments	Sources
1831 BC		Xia China *see* Mount Tai earthquake			?	?	Listed in the *Bamboo Annals*	[1][2]
464 BC		Sparta, Greece *see* 464 BC Sparta earthquake			?	7.2 (approx)	M_s Led to a helot uprising and strained relations with Athens, one of the factors that led to the Peloponnesian War	[3]
226 BC		Rhodes, Greece *see* 226 BC Rhodes earthquake			?	–	Destroyed Colossus of Rhodes and city of Kameiros	[4]
60 BC		Portugal and Galicia coasts			?	8.5	Caused a tsunami	[5]
17 AD	At night	Asia minor *see* 17 AD Lydia earthquake	37.85	27.3	?		Destroyed 13 cities in Asia (minor)	Described by the historians Tacitus and Pliny the Elder
February 5, 62		Bay of Naples, Italy *see* 62 Pompeii earthquake			?	5–6	Brought down a large part of Pompeii, caused severe damage in Herculaneum and Nuceria.	Seneca describes it in his "Quaestiones Naturales VI" [6]
110		Dian Kingdom, Yunnan, southwestern China			probably thousands	-	Flooded administrative centre of the Dian Kingdom	[7]
December 13, 115		Antioch *see* 115 Antioch earthquake	36.1	36.1	~260,000	7.5	M_s	[8]
May 18, 363		Syria *see* 363 earthquake			"thousands"	~7	Destruction also in "The Holy Land", Petra	Ammianus Marcellinus[9] and numerous other late Antiquity writers[10]
July 21, 365		Crete (Greece) *see* 365 Crete earthquake			"thousands"	XI	Destruction also in Cyrene & Alexandria (by tsunami). Uplifted Crete by 9 metres.	Ammianus Marcellinus[9] and numerous other late Antiquity writers[10]

382		Cape St. Vincent, Portugal				7.5	According to Roman historian Ammianus Marcellinus, the earthquake and corresponding tsunami sank two islets that were situated near Cape St. Vincent.	Ammianus Marcellinus
May 19, 526		Antioch, Turkey *see* 526 Antioch earthquake			250,000	VIII	The city of Antioch was greatly damaged, and some decades later the city's population was just 300,000.	Procopius, II.14.6; sources based on John of Ephesus
July 6, 551		Beirut, Tyre, Tripoli *see* 551 Beirut earthquake	33.9	35.5	30,000	7.5	M_w Triggered a devastating tsunami, all the cities of the Phoenician coast from Tyre to Tripoli were reduced to ruins	[11]
January 18, 749[12][13]		The Levant *see* 749 Galilee earthquake				7 to 7.5 (approx)	The cities of Tiberias, Beit She'an, Hippos and Pella were largely destroyed while many other cities across the Levant region were heavily damaged.	[12]
November 24, 847		Damascus, Syria *see* 847 Antioch earthquake	33.5	36.3	70,000	X		[11][14]
December 856		Corinth, Greece	37.9	22.9	45,000	–		[14][15]
December 22, 856 (aftershocks for about a year)		Qumis, Iran. From Khuvar to Bustam and Gurgan. The town of Damghan hardest hit. *see* 856 Damghan earthquake	36.23	54.14	45,000–200,000.		The city of Damghan was half destroyed and had 45,096 casualties.	[14][16]
July 13, 869		Sendai, Japan *see* 869 Jogan Sanriku earthquake	38.5	143.8	~1,000	8.6–9.0	M_s	[17]

March 23, 893		Ardabil, Iran *see* 893 Ardabil earthquake	38.28	48.30	150,000	–	–	Regarded as a 'fake earthquake', due to misunderstanding of original Armenian sources for the 893 Dvin event.[18][19][20][21]
December 28, 893		Dvin, Armenia *see* 893 Dvin earthquake	40.0	44.6	30,000	–	Mislocated in India	[22]

11th–18th centuries

Date	Time‡	Place	Latitude	Longitude	Fatalities	Magnitude	Comments	Sources
December, 1037		Taizhou, Jiangsu, China	32.0	119.0	22,391			[23][24]
August 12, 1042		Palmyra, Baalbek, Syria	35.1	38.9	50,000	7.2 (>VIII)		[11][14]
March 18, 1068		Near East *see* 1068 Near East earthquake			20,000	≥ 7.0		[25][26]
October 11, 1138		Aleppo, Syria *see* 1138 Aleppo earthquake	36.1	36.8	230,000	XI		[11][27]
August 12, 1157	08:15	Hama, Syria *see* 1157 Hama earthquake	35.1	36.3	"Tens of thousands"	7.2	M_s Largest in a sequence lasting from late 1156 to early 1159	[11][27][28]
February 4, 1169		Sicily *see* 1169 Sicily earthquake	37.3	15.0	15,000	X		[29]
June 29, 1170	06:29	Eastern Mediterranean	34.4	36.4	5,000[30]-80,000 in Aleppo 25,000 in Hama	7.3–7.5[31]-7.7[32]	Syria, Lebanon, central southern Turkey	Numerous sources from Crusader times[27][28]
July 5, 1201 and/or May 20, 1202		Eastern Mediterranean; see 1202 Syria earthquake			1,100,000 (includes famine/ disease deaths)	7.6	Damage across a wide area from Syria to Upper Egypt	–
May 11, 1222	06:15	Cyprus *see* 1222 Cyprus earthquake	34.7	32.6		7.0–7.5	Caused damage at Paphos, Limassol and Nicosia	[33]
1268		Cilicia, Anatolia (Armenian Kingdom of Cilicia) *see* 1268 Cilicia earthquake	37.5	35.5	60,000	7 (approx)		–
September 27, 1290		Chihli (Hopeh), China *see* 1290 Chihli earthquake	41.5	119.3	100,000	6.8	M_s	[34]

Date	Time‡	Place	Latitude	Longitude	Fatalities	Magnitude	Comments	Sources
May 26, 1293		Kamakura, (now in Kanagawa Prefecture, Kantō region,) Japan	35.2	139.4	23,024	7.1		[35]
August 8, 1303, 06:00		1303 Crete earthquake, Greece	35	27	up to 10,000	~8	Triggered a major tsunami that devastated Alexandria in Egypt	[36]
January 1, 1341		Crimea, Ukraine			Not known	6		[37]
January 25, 1348	15:00	Friuli, Venice, Rome *see* Earthquake of 1348	46.37	13.58	10,000	6.9		[14]
October 18, 1356		Basel, Switzerland *see* 1356 Basel earthquake	47.5	07.6	1,000	6.2	M_w	[38]
May 21, 1382		Canterbury, UK			?	5.8	Struck during synod – later called "Earthquake Synod" – called to condemn heresy of John Wycliffe – some saw as portentous	[39][40]
February 2, 1428		Catalonia (now Spain) Catalan earthquake of 1428	42.4	2.2	1,000s	VIII–IX	Sometimes called the *terratrèmol de la candelera* because it took place during the Candlemas.	[14][41][42]
3 May 1481	03:00	Rhodes *see* 1481 Rhodes earthquake	36.0	28.0	30,000	7.1	Largest of a series that lasted 10 months	[43]
September 20, 1498	08:00 local time	Honshu *see* 1498 Meiō Nankaidō earthquake	34.0	138.1	31,000	8.6	M_s	[44]
June 6, 1505		Nepal *see* 1505 Lo Mustang earthquake	29.5	83.0	30% of Nepalese population	8.2		-
September 10, 1509	22:00	Istanbul, Turkey *see* 1509 Istanbul earthquake	40.9	28.7	10,000	7.2	M_s	[45]
January 26, 1531	04:30	Lisbon, Portugal *see* 1531 Lisbon earthquake	38.9	-09.0	30,000	6.9		[46]
January 23, 1556		Shaanxi, China *see* 1556 Shaanxi earthquake	34.5	109.7	830,000+	8.2–8.3	Deadliest earthquake in recorded history	USGS
November 16–17, 1570	19:10	Ferrara, Italy *see* 1570 Ferrara earthquake	44.817	11.633	70–200	5.5		Azariah de Rossi's *Kol Elohim*[14][47]
December 16, 1575	18:30	Valdivia, Chile *see* 1575 Valdivia earthquake	-39.8	-73.2	?	8.5		[14][48]
November 24, 1604	12:30 local time	Arica, Chile	-18.500	-70.400	?	8.5	M_l	[49]
February 3, 1605	20:00 local time	Shikoku, Honshu, Japan *see* 1605 Keichō Nankaido earthquake	33.5	138.5	thousands	7.9	M_s	[50]

Date	Time‡	Place	Latitude	Longitude	Fatalities	Magnitude	Comments	Sources
July 13, 1605		Qiongshan, Hainan, China	19.9	110.5	3,000	7.5 (X)		[51]
February 5, 1663		Quebec, Canada *see* 1663 Charlevoix earthquake	47.6	70.1	0	7.3–7.9	Landslides were the primary feature	[52]
April 6, 1667		Dubrovnik, Croatia	42.3	18.1	3,000	7.2		[53]
November 25, 1667		Shamakhi, Azerbaijan *see* 1667 Shamakhi earthquake	40.6	48.6	80,000	6.9	M_s	[34]
August 17, 1668		Anatolia, Turkey	40	36	8,000	8		USGS
October 20, 1687	11:30	Lima, Peru *see* 1687 Peru earthquake	-15.2	-75.9	5,000	8.2		[14]
September 13, 1692	11:00	Salta Province, Argentina *see* 1692 Salta earthquake	-25.40	-64.80	~13	7.0	The small village of Talavera del Esteco was completely destroyed.	[54]
June 7, 1692	11:43 local time	Port Royal, Jamaica *see* 1692 Jamaica earthquake	17.9	-76.8	2,000+	7 (approx) (X)		[55][56]
January 11, 1693		Catania Province, Sicily *see* 1693 Sicily earthquake			60,000	7.5		–
September 5, 1694	11:40	Irpinia, Italy *see* 1694 Irpinia–Basilicata earthquake	40.88	15.35	6,000	6.9	M_w	[57]
January 26, 1700	~21:00	Cascadia subduction zone *see* 1700 Cascadia earthquake				9	M (Satake et al., 1996)	USGS
January 14, 1703	18:00	Norcia, Italy *see* 1703 Apennine earthquakes	42.7	13.07	6,240–9,761	6.7	M_w	[14][58]
February 2, 1703	11:05	L'Aquila, Italy *see* 1703 Apennine earthquakes	42.43	13.3	2,500–5,000	6.7	M_w	[14][58]
December 31, 1703	17:00	Kanto region, Japan *see* 1703 Genroku earthquake	35.0	140.0	5,233	8.2	M_s	[59]
October 28, 1707	14:00 local time	Japan *see* 1707 Hōei earthquake	33.0	136.0	5,000+	8.6		[60]
November 18, 1727		Tabriz, Iran *see* 1727 Tabriz earthquake	38.0	46.3	77,000	VIII		[61]
July 8, 1730	08:45	Valparaiso, Chile *see* 1730 Valparaiso earthquake	-32.5	-71.5	?	8.7		[62]
September 30, 1730	10:00 local time	Beijing, China	40.0	116.2	100s	6.5		[63][64]

Date	Time‡	Place	Latitude	Longitude	Fatalities	Magnitude	Comments	Sources
October 16, 1737	15:30 local time	Kamchatka, Russia	51.1	158.0		8.3	M_s See Kamchatka earthquakes	[65][66]
October 28, 1746	22:30 local time	Lima & Callao, Peru *see* 1746 Lima–Callao earthquake	-11.35	-77.28	4,000–5,000	8.6–8.8	M_w	[67]
May 25, 1751	1:00 local time	Concepción, Chile *see* 1751 Concepción earthquake	-36.830	-73.030		8.5	M_l	USGS
June 7, 1755		Northern Persia	34.0	51.5	1,200 40,000[68]	5.9		[64][69]
November 1, 1755	10:16	Lisbon, Portugal *see* 1755 Lisbon earthquake	36	-11	80,000	8.7	Caused a huge tsunami	USGS
November 18, 1755	09:11	Boston, Massachusetts, USA *see* 1755 Cape Ann earthquake	42.7	-70.2	0	5.9	M_w	[70][71]
November 25, 1759	19:23 local time	Eastern Mediterranean *see* Near East earthquakes of 1759	33.7	35.9	1,000s	7.4	M_s Earthquake in same area on October 30 considered to be a foreshock	[11]
April 2, 1762		Northeastern Bay of Bengal *see* 1762 Arakan earthquake	22.0	92.0	200	up to 8.8	M_w	[72]
June 28, 1763	05:28	Komárom, Kingdom of Hungary *see* 1763 Komárom earthquake	47.73	18.15	83	6.2 to 6.5	M_w	[73]
October 21, 1766		Saint Joseph, Trinidad and Tobago	10.65	-61.43		7.9	Destroyed Spanish colonial capital of San Jose, Trinidad (now St. Joseph)	[74]
June 3, 1770	19:15 local time	Port-au-Prince, Haiti	18.7	-72.63	200+	7.5	M_w	[75]
July 29, 1773		Guatemala *see* 1773 Guatemala earthquake	14.6	-90.7	5–600	7.5		
January 8, 1780		Iran	38.0	46.2	200,000	7.4	M_s	[76][77]
February 4—5, March 28, 1783	12:00	Calabria, Italy *see* 1783 Calabrian earthquakes	38.15	15.70	35,000	6.9	M_w First in a sequence of five earthquakes $M_w \geq 5.9$ to hit Calabria in less than two months	[78]
June 1, 1786	04:00 local time	Sichuan, China *see* 1786 Kangding-Luding earthquake	29.9	102.0	~100,000	7.75	M_w Triggered a landslide that blocked the Dadu river – the collapse of the dam during an aftershock and subsequent flood caused most of the casualties	[79]
February 4, 1797	12:30	Quito, Ecuador & Cuzco, Peru *see* 1797 Riobamba earthquake			41,000	7.3	M_w	[80]

Date	Time‡	Place	Latitude	Longitude	Fatalities	Magnitude	Comments	Sources
February 10, 1797		Sumatra, East Indies (now Indonesia) *see* 1797 Sumatra earthquake	-1.0	99.0	300	8.4		[81][82][83]
November 22, 1800	1:30 p.m. local time	San Diego, California, USA	34.00	-117.18	Unknown	6.5	Resulted in varying levels of damage at Mission San Diego de Alcalá, Mission San Luis Rey de Francia, and Mission San Juan Capistrano	[84]

19th century

Date	Time‡	Place	Latitude	Longitude	Fatalities	Magnitude	Comments	Sources
October 26, 1802	10:55	Vrancea region, Moldavia, now Romania	45.7	26.6	3 in Bucharest	7.9	Serious damage in the area. All church steeples in Bucharest collapsed, as well as many houses and Colțea bell tower.	[85][86]
February 16, 1810	22:15	Crete, Heraklion *see* 1810 Crete earthquake	35.5	25.6	2,000	7.5	M_w	[87]
December 16, 1811	08:00	New Madrid, Missouri, USA *see* 1811–12 New Madrid earthquakes	36.6	-89.6		8.1	M_l (Johnston, 1996)	USGS
January 23, 1812	15:00	New Madrid, Missouri, USA *see* 1811–12 New Madrid earthquakes	36.6	-89.6		7.8	M_l (Johnston, 1996)	USGS
February 7, 1812	09:45	New Madrid, Missouri, USA *see* 1811–12 New Madrid earthquakes	36.6	-89.6		8	M_l (Johnston, 1996)	USGS
December 8, 1812	9:45 a.m. local time	Wrightwood, California, USA*see* 1812 Wrightwood earthquake	34.22	-117.39	42	7.0	Destroyed the “Great Stone Church” at Mission San Juan Capistrano	Southern California Earthquake Data Center

Date	Time‡	Place	Latitude	Longitude	Fatalities	Magnitude	Comments	Sources
December 21, 1812	11:00 a.m. local time	Santa Barbara Channel, California, USA	34.12	-119.54	0	7.0	Destroyed the church at Mission Santa Barbara, caused near-total destruction at Mission La Purísima Concepción, and considerable damage at Mission Santa Inés	Southern California Earthquake Data Center
June 16, 1819	18:45–18:50 local time	Gujarat, India *see* 1819 Rann of Kutch earthquake	23.0	71.0	>1,543	7.7–8.2	M_w Formed an 80 km long ridge, the Allah Bund ('Dam of God')	[88]
June 2, 1823	08:00	south flank of Kīlauea, Hawaii, Kingdom of Hawaii	19.3	-155		7	M_l (Klein and Wright, 2000)	USGS
November 25, 1833		Sumatra, East Indies (now Indonesia) *see* 1833 Sumatra earthquake	-2.5	100.5	"numerous victims"	8.8–9.2	M_w Huge tsunami flooded all southern part of western Sumatra	[81][83][89]
June 10, 1836	15:30	south San Francisco Bay region, California, Mexico (now USA)	36.9	-121.5		6.5	M_l (Bakun, 1999)	USGS
January 1, 1837	16:00 local time	Galilee, Palestine *see* Galilee earthquake of 1837	33.0	35.5	6,000–7,000	>7.0	M_s	[11][90]
June 1838		San Francisco Peninsula, California, Mexico (now USA)	37.3	-123.2		6.8	M_l (Bakun, 1999)	USGS
January 5, 1843	02:45	Marked Tree, Arkansas, USA	35.5	-90.5		6.3	M_l (Johnston, 1996)	USGS
May 8, 1847		Around Zenkō-ji Temple, (now in Nagano Prefecture, Chūbu region,) Japan	36.7	138.2	more than 8,600	7.4	2,094 houses lost by fire with quake at Nagano	[91]
23 December 1854	09:00	Honshu, Japan *see* 1854 Tōkai earthquake	34.0	137.8	2,000	8.4	Major tsunami	[92]

Date	Time‡	Place	Latitude	Longitude	Fatalities	Magnitude	Comments	Sources
24 December 1854	16:00	Honshu, Japan *see* 1854 Nankai earthquake	33.0	135.0	thousands	8.4	Major tsunami	[92]
January 23, 1855	21:11 local time	Wairarapa, New Zealand *see* 1855 Wairarapa earthquake	-41.4	174.5	4	8.0 (approx)	Raised sections of Wellington coastline by 2 metres	–
November 11, 1855	22:00 local time	Edo (now Tokyo, Kantō region), Japan *see* 1855 Ansei Edo earthquake	35.65	139.8	~ 7,000	7.0	M_s 10,000 house destroyed in Edo	安政の大地震 in Japanese
January 9, 1857	16:24	Fort Tejon, California, USA (San Andreas fault from Parkfield to Wrightwood) *see* 1857 Fort Tejon earthquake			1	7.9	M (Grant and Sieh, 1993; Stein and Hanks, 1998)	USGS
December 16, 1857	21:00	Naples, Italy *see* 1857 Basilicata earthquake	40.3	16	11,000	6.9	M_I	USGS
February 16, 1861		Sumatra, East Indies (now Indonesia) *see* 1861 Sumatra earthquake	1.0	97,5	905	8.5	Caused major tsunami.	[93][94]
October 8, 1865	20:46	San Jose, California, USA	37.2	-121.9		6.5	M_I (Bakun, 1999)	USGS
April 3, 1868	02:25	Kau, southeast Hawaii, Hawaii, Kingdom of Hawaii *see* 1868 Hawaii earthquake	19.2	-155.5	77	7.9	M_I (Klein and Wright, 2000)	USGS
August 13, 1868	16:45 local time	Arica, Chile *see* 1868 Arica earthquake	-18.500	-70.350	25,000	9.0	M_I	USGS
October 21, 1868	15:53	Hayward, California, USA, Hayward Fault Zone *see* 1868 Hayward earthquake	37.7	-122.1	30	6.8	M_I (Bakun, 1999)	USGS
February 20, 1871	08:42	Molokai, Hawaii, Kingdom of Hawaii	21.2	-156.9		6.8	M_I (Klein and Wright, 2000)	USGS

Date	Time‡	Place	Latitude	Longitude	Fatalities	Magnitude	Comments	Sources
March 26, 1872	10:30	Owens Valley, California, USA *see* 1872 Lone Pine earthquake	36.5	-118	27	7.6	M (Beanland and Clark, 1994)	USGS
December 15, 1872	05:40	North Cascades, Washington, USA *see* 1872 North Cascades earthquake	47.9	-120.3		7.3	M_I (Malone and Bor, 1979; Rogers et al., 1983)	USGS
May 10, 1877	21:16 local time	Iquique, Chile *see* 1877 Iquique earthquake	-19.600	-70.230	2,541	8.8	M_w	USGS
September 9, 1880	7:04 local time	Zagreb, Croatia *see* 1880 Zagreb earthquake	45.9	16.1	1	6.2	M_s	[95]
3 April 1881	11:30	Chios, Çeşme, Alaçatı *see* 1881 Chios earthquake	38.25	26.25	7,866	6.5	M_w	[96]
December 31, 1881	01:49	India, Andaman and Nicobar Islands *see* 1881 Nicobar Islands earthquake	8.52	92.43	0	7.9	M_w	[97]
August 31, 1886	02:51	Charleston, South Carolina, USA *see* 1886 Charleston earthquake	32.9	-80	60	7.3	M_I (Johnston, 1996)	USGS
September 1, 1888	04:10 local time	North Canterbury, New Zealand *see* 1888 North Canterbury earthquake	-42.6	172.4		7.0–7.3	First earthquake observed to be associated with mainly horizontal fault displacement	–
July 28, 1889	23:40	Kumamoto Prefecture, Japan *see* 1889 Kumamoto earthquake	32.8	130.7	20	6.3		[14]
October 27, 1891	21:38	Mino-Owari (Gifu-Aichi Prefectures), Tōkai region, Japan *see* Nōbi earthquake	35.6	136.6	7,273	8	M_S	USGS
April 19, 1892	10:50	Vacaville, California, USA	38.5	-121.8	1	6.4	M_I (Bakun, 1999)	USGS
April 21, 1892	17:43	Winters, California, USA	38.6	-122		6.4	M_I (Bakun, 1999)	USGS

Date	Time‡	Place	Latitude	**Longitude**	**Fatalities**	**Magnitude**	**Comments**	**Sources**
November 17, 1893	15:06	Quchan, Iran *see* 1893 Quchan earthquake	37.2	58.4	18,000	6.6	M_s	[98]
October 31, 1895	11:08	Charleston, Missouri, USA	37	-89.4		6.6	M_I (Johnston, 1996)	USGS
June 15, 1896	19:32	Off the Pacific coast of Sanriku, Iwate Prefecture, Tōhoku region, Japan *see* 1896 Sanriku earthquake	39.5	144	22,000+	8.0–8.1 (approx)	M_w	USGS
June 12, 1897	11:06	Assam, India *see* 1897 Assam earthquake	26	91	1,500	8.3		USGS
September 21, 1897	05:12	Mindanao, Philippines *see* 1897 Mindanao earthquakes	6.0	122.0	13	7.5	M_S	[99][100]
September 4, 1899	00:22	Cape Yakataga, Alaska, USA	60	-142		7.9	M_S	USGS
September 10, 1899	21:41	Yakutat Bay, Alaska, USA	60	-142		8	M_S	USGS
October 9, 1900	12:28	Kodiak Island, Alaska, USA	57.1	-153.5		7.7	M_S	USGS

CHAPTER 7

SOUND WAVES

INTRODUCTION

People communicate through their senses. Communication is the broadcast and reception of meaningful information. Usually people do communicate with each other within the sense of sight and hearing through visual and vocal means, therefore by light and sound. Through different technological devices such as telephone, telegraph, radio, radio telescope, orbiting satellites, microwave relays and lasers as well as the development applied in the field of electronics, communication in human race is the beneficiary of unlimited favors. Even today, the hindrance between time and the distance prohibited efficient and fast interaction. The communication through sounds allows people from opposites places talk and interact as if they are just in the same place. The conversion of light and sound waves to electrical energy, and the utilization of electromagnetic waves for the broadcast and reception of information.

Sound waves are very important part of communication. Without sound waves, human beings are impossible to communicate verbally. The vocal chords produce sound waves that are conveyed through the air to the ears of the receivers of information. Modern communication technologies like radio and television use common concepts in transmitting sound to human ears.

Sonar devices which utilize sound waves that are used by the scientists in exploring the oceans. Sound waves are being send-off by sonar which will be bounced back to the origin when they hit an object. The echo propagated by this equipment is used by the scientists to know the size and the distance of the object that rebound the sound waves back to its origin. Navy vessels use the sonar technology in locating the enemy's submarines. Sound waves are also used by geologists in searching oil underneath the earth. They rebound sound waves into the ground and calculate the way in which they transport through the earth. Density references were made by geologists in order to measure how sound waves travel.

Sound waves are also used by many creatures to hunt for their food. Bats use sonar to search for food. Bats emit sound waves that bounce off prey. They can determine the distance of their prey from their location when the sound waves are bounced back to them. Sea dolphins also use the same form of sonar or echolocation to both, to hunt for prey and to communicate with one another.

NATURE OF SOUND WAVES

Telephone which is an electronic vocal communication device require the mutual conversion of sound and electricity. People live in a world of sound. Everywhere they go and whatever they do, they hear sounds. Some sounds are very familiar, a friend's voice, the chirping of the birds, the ticking of the clock, the barking of the dogs and the beating of your heart. Certain sounds like music are pleasant to the ears; others are not which are categorized as noise. Different sounds have different effects. For example, music calms and relaxes. The crashing sound of explosives or thunderous sound made by jet when it takes off can make us feel nervous.

Sound is confined to the frequency range of 20 Hz to 20, 000 Hz to which the human ear is sensitive. Waves with frequencies below this audible range are called infrasound and those above are referred to as ultrasounds.

Sound waves are longitudinal waves. They are produced by a sequence of vibrations parallel to the direction of travel of the waves. When the strings of guitar was plucked, the strings look blurred as they tremble. You can feel the vibrations of your throat as you touch it when talking; that vibrations are the trembling of your vocal chords. In ringing a bell, pull the string and move it back and forth to generate sound. All these examples, the source of the sound waves is the vibrating objects.

Mechanical Sound Waves

A wave can be characterized as a vibration that travels through a medium, transmitting energy from one direction to another direction. The medium is commonly an object through which the vibration is moving; it can be perceived of as a sequence of cooperating particles. The example of a slinky wave is often used to describe the nature of a wave. A vibration is usually propagated within the slinky by the back and forth movement of the first coil of the slinky. The first coil becomes strained and starts to push and pull on the succeeding coil. The second coil will be disturbed from its equilibrium state because of the pull and push on the second coil. The second coil moves, it starts to push and pull on the third coil; the push or pull on the third coil moved its equilibrium state. As the third coil becomes expatriate, it starts to push and pull to the fourth coil. This process continues in succeeding coils with each separate particle acting to move the adjacent particle. Then the vibrations transport through the slinky. As the disturbance moves from coil to coil, the energy that was originally introduced into the first coil is transported along the medium from one location to another.

A sound wave is the same in nature regarding the slinky wave for some reasons. Reason number one, there is medium that carries the vibrations from one location to another. Naturally, the medium for sound to travel is the air, though sound can also be transmitted through water or steel. The medium is defined as the series of interconnected and interacting particles. Second, the original source of the wave is present, some objects that are vibrating capable of displacing the first particle of the medium. The vibration could be propagated by the vibrating vocal chords of a person, the vibrating string and soundboard of a guitar or violin, the disturbance created by tuning fork, or the vibrating diaphragm of a radio speaker. Third reason the sound wave is transmitted from one location to another by means of particle to particle interaction. One particle of the air is moved from its equilibrium when the air is moving through the air; it employs push and pull to the object nearest to it causing them as well to displace from their equilibrium state. The interaction of this particle continues throughout the whole medium, each particle interacting and causing vibrations to its nearest neighbours. Sound wave is categorized as mechanical wave because sound wave is a vibration that is transmitted through a medium via the way or particle-to-particle interaction.

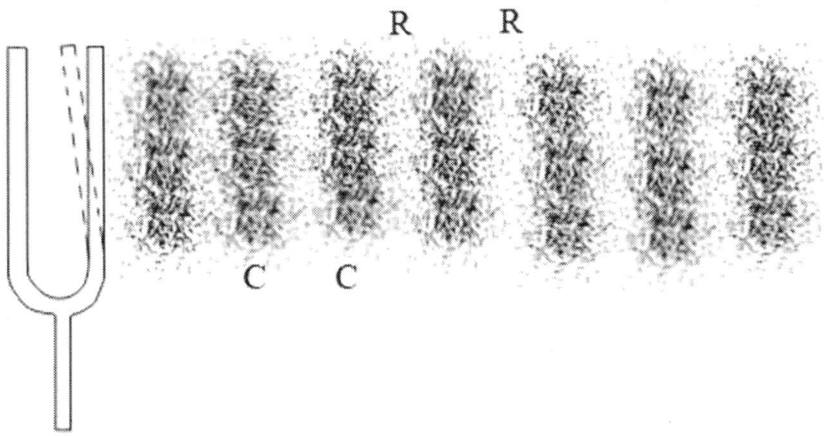

Figure 7.1: The vibrations caused in the medium due to its propagation are known as wave motion.

Sound waves are categorized as mechanical waves.

Longitudinal Sound Waves

The waves which motion of individual particles of the medium is parallel to the direction of energy transport are called longitudinal waves. It can be propagated from the slinky if it is stretched out in a horizontal direction and the first coil of the slinky are vibrated horizontally. In that case, each coil of the medium will vibrate into the direction parallel to the energy that is being transported.

Sound waves in air can be categorized as longitudinal waves because particles of the medium which sound is transmitted vibrate parallel to the direction of the movement of sound wave.

A string that vibrates can propagated longitudinal waves. When the vibrating string is displaced forward direction, then it starts to push air on the surrounding molecules, displacing them toward their nearest neighbors. This may cause the air molecules to the right of the string to be compacted into a small region of space. The air molecules will move back leftward as the string that vibrates moves in the opposite direction and lowers the pressure of the air. Due to the lower pressure to the right of the string, air molecules in that portion directly to the right of the string to expand into large portion of space. When the string moves back and forth that causes vibration, the air molecules continually vibrate back and forth horizontally to the right of the string. The molecules or air move rightward as the strings move rightward and move leftward as the strings move leftward. Therefore, the back and forth vibrations are communicated to nearest neighbors by the interaction called particle-to-particle. Other particles in the surrounding start to displace rightward and leftward, therefore transfer a wave to the right. Due to the parallel movement of the air molecules to the direction of the wave moves, the sound wave is referred to as a longitudinal wave. Compressions and rarefactions within the air are created when longitudinal vibrations on the air is resulted from such phenomenon.

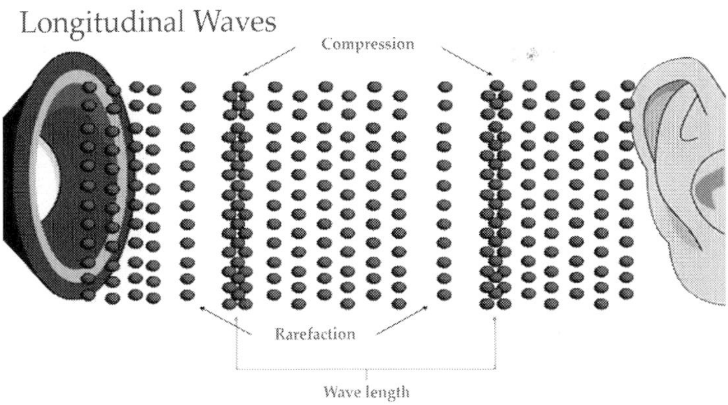

Figure 7.2: Sound waves as longitudinal waves.

Sound as Pressure Waves

Sound is a mechanical wave that outcomes from the back and forth disturbance of the particles of the medium wherein the sound wave is transmitting. When a sound wave is transmitting from left to right via air, therefore the particles of the air will be transferred toward the right and left direction while the energy of the sound waves travels on it. Sound waves are categorized as longitudinal waves because the movement of particles of the medium is parallel to the direction of the energy that the sound waves carry.

Difference between Compressions and Rarefactions

A longitudinal wave is best created through a pulsating tuning fork. When the prongs of the fork pulsate back and forth, they shove air particles that are near. As the prongs moves to the forward motion it shove the molecules of the air to the right and regressive retraction of the prong generates a low-pressure area permitting the particles of the air to transfer from back to the left. Due to the longitudinal movement of particles of the air, there are parts in the where the particles of the air are compacted together and other parts where the particles of the air are separated with gaps. These parts are called as compressions and rarefactions correspondingly. High pressure corresponds compressions and parts with low pressure corresponds to rarefaction.

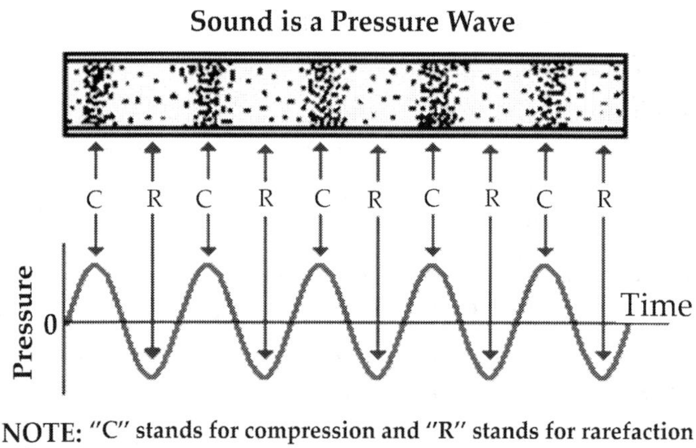

Figure 7.3: Sound wave is a pressure wave.

The illustration below portrays a sound wave that is produced by tuning fork and proliferated through air in an open tube. It also shows the compressions and rarefactions created by the tuning fork.

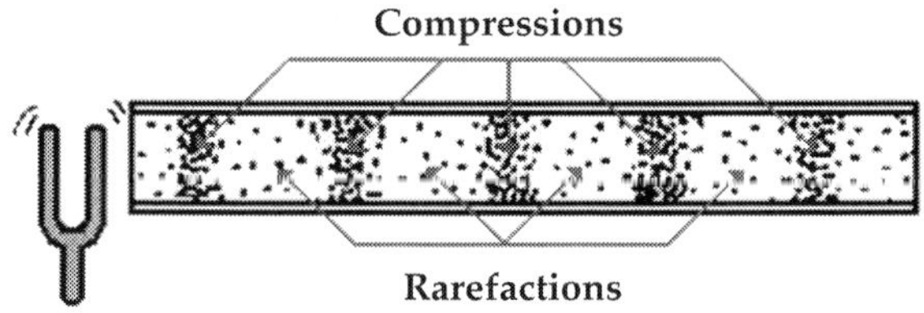

Figure 7.4: Show the compressions and rarefaction produce by the tuning fork.

The wave's wavelength is simply the distance that a vibration transmits through a media in a complete wave cycle. Once every wave cycle a wave repeats its cycle, the wavelengths are seldom reiterating arrangement, the length of one complete wave. In the case of transverse wave, this distance is usually calculated starting from one wave crest to the next nearest wave crest or from one wave, trough to the next nearest trough. It's just that longitudinal wave lack of crests and troughs, its wavelengths must be subjected to differently calculations. Repeating pattern of compressions and rarefactions comprises longitudinal wave. Therefore, the wavelength is usually calculated as the distance form one compression to the nearest compression or the space to the nearest compression or the space from rarefaction.

Pressure Wave

Sound wave is composed of reiterating arrangement of high-pressure and low-pressure parts transferring through a medium, the reason why it is also regarded as a pressure wave. In using a detector such as human ear or a man-made instrument, were be utilized to distinguish a sound wave. It senses oscillations in pressure as the sound wave interrupts upon sensing the detecting device. At a very reliable time, the detector would perceive a high pressure; this resembles the influx of a compression at the detector site. The detector might sense a normal pressure in the next prompt in time. Finally the low pressure will be sensed normal pressure as the coming of a rarefaction at the site of detector. The fluxes in pressure as sensed by the detector appear at periodic and consistent intervals in time. Actually a sine curve will show in a plot of pressure against time. Compressions correspond to the highest points of the sine curve; rarefactions resemble the low points; and the pressure that the air would consist if there were no vibrations travelling on it is called the zero points. Figure 7.3 might lead to a misconception if it is not carefully comprehend. The illustration of sound by a sine wave is simply an effort to demonstrate the sinusoidal nature of the pressure-time oscillations. It must not be perceived that sound is a transverse wave that are composed of crest and troughs. Sound waves are longitudinal waves that transmit through air consisting of compressions and rarefactions. The air particles do not tremble as the sound travels through it or in any medium that is fluid. Sound waves that are conveying through air are longitudinal waves.

PROPERTIES OF SOUND

The Human Ear

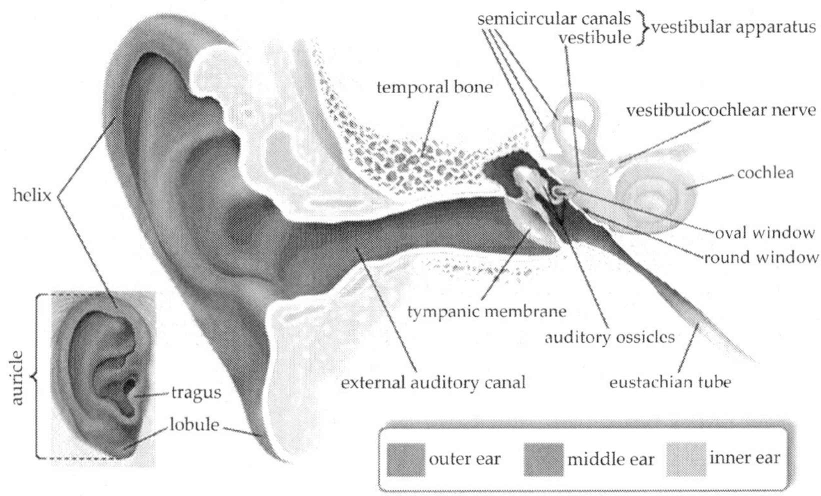

Figure 7.4: The human ear.

Sound waves are detected by the ears. Human ears are the organs of hearing and balance that perceive and examine noises through converting sound waves to electrochemical impulses and retains the sense of balance. There are two various functions of that the human ears and other mammals comprise, the hearing and equilibrium, and the harmonization of head and eye activities. The outer ear, middle, and inner ear are the three distinct parts of the ear. The outer is composed of pinna, external auditory canal and tympanic membrane also known as eardrum. Pinna or auricle is the visible portion of the ear that can be found at the side of the head. Pinna collects the sound then pass through the external auditory canal then lead the sound waves to the tympanic membrane called eardrum.

The middle ear is a contracted, air-filled cavity in the temporal bone. It is covered by a series of three bones the malleus also known as the hammer, incus which is known as anvil and the stapes also called as the stirrup, they are collectively known as the auditory ossicles. This sequence of bones connect the sound collected from the tympanic membrane to the inner ear, which is also known as labyrinth. The term labyrinth was already proposed on time of Galen. It is a complex system of passages filled with fluid and cavities that can be found deep inside the petrous part of the temporal bone.

The inner ear is composed of two units that are functional, the vestibular apparatus and cochlea. Vestibular apparatus comprising of the vestibule and canals that are semi-circular which enclose the sensory organ of the postural equilibrium. Meanwhile, cochlea assumes a snail-shell shape, comprises the sensory organ of hearing. These sensory organs are extremely specialized endings of eighth cranial nerve also known as vestibulocochlear nerve.

The nerves at the hair cells transmit the message or the sound waves to the hearing center of the brain. The sound memory center of the brain stores messages, which is its function since the human birth, and then identify the sounds received. People who have hearing impairment do not use their ears to sense sound. Other parts of their body are used to perceive sound. A famous inventor who had hearing impairment would listen to phonograph records by resting his head on the phonograph player. The vibrations created by the phonograph transmit through the bones of his head. Others can play the violin even though they have partially impaired hearing through resting their chin on the instrument. The disturbances created by the violin will be conveyed through the jaw bone of the player.

Pitch and Frequency of Sound Waves

Some sounds perceive to be low and others are perceived to be high. Pitch is the highness or lowness of the sound. Pitch is a characteristic of sound that can be detected by our senses. The frequency of the sound waves govern the pitch of the sound. The higher the frequency of a sound wave, the higher the pitch.

Human beings have confined hearing capacity. The eardrums of the person can create a disturbance only within a specific range of frequency. The human hearing capacity ranges from 20 Hz to 20, 000 Hz. Frequency outside of this range, vibrations cannot be detected by human. Beyond 20, 000 Hz, a person may hear sound but there will a difficulty in identifying the difference in pitch. Ultrasounds or waves with frequencies above 20, 000 Hz is already hazardous to human, there will be bad effects once a person is exposed to this kind of frequency. 20, 000 Hz is the upper limit of human hearing. Below 20 Hz, the frequencies of the infrasound is the lower limit of hearing, only vibrations will be felt.

People has the ability to change the pitch of their voice by regulating the opening of their vocal cords. When the opening of the vocal cord is small, the pitch is high, when it is wide, the pitch is low. In order to change the voice pitch you must change the shape of the mouth as well.

Musical notes have different frequencies. The pitch transformed every time that the player needs to play a specific note. For an example, the musician needs the note re from his or her instrument, he she has to change the pitch to find re in the musical scale.

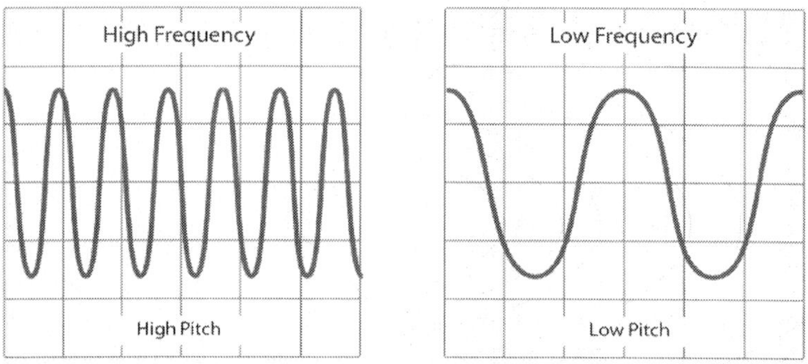

Figure 7.5: shows the relationship between the frequency and the pitch of the sound waves.

Intensity and Loudness

Sound intensity is the quantity of energy carried by sound while loudness is the calculation of the sound audible to the receiver. The idea about sound intensity and loudness are essential in the field of music, audio engineering, physics and other variety of fields. Sound intensity is a quantity of the energy of a sound wave and refers to the specific physical calculation. It is also perceived as the energy carried by a wave in an area per unit of time. The intensity differs as the amplitude of wave becomes greater. Therefore, the greater the variations between the compressions and rarefaction of the sound wave, the higher its intensity. Intensity is nearly associated to more subjective word loudness. Loudness refers to how the sound is perceived by the receiver. For an instance, a sound that gives an impression that it is loud in quiet room would probably be unheard when it is uttered or played in a busy street corner, even though the sound has the same intensity.

Moreover, loudness is characterized as the quality of hearing sensation in a way of which sounds can be determined on a scale ranging from quiet to loud. Loudness may vary on few qualities of sound such as amplitude, frequency and duration.

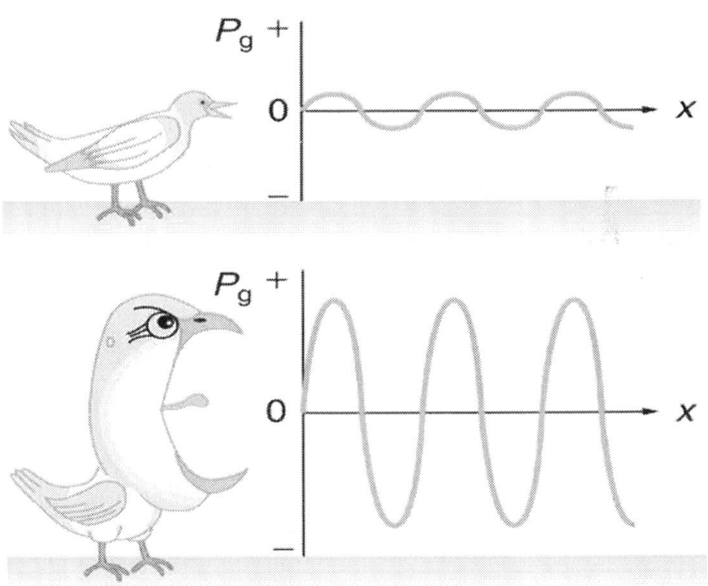

Figure 7.6: shows the relationship between sound intensity and loudness.

Sound waves always carry energy. A pulsating body such as guitar string or the drumhead, generates sounds when a great amount of energy is conveyed on it. Sound waves transport energy through the medium's particles. Soft sounds have lesser amplitude compared to loud sounds. Sound waves transmit in all directions. But through a megaphone, the sound wave is focused by the walls of the device. This strengthens the sound wave in one direction and produces a very loud sound. Loud sounds have greater amplitude therefore the energy transmitted to the medium's particles is greater. In observing and sensing sounds inside the body, doctors utilize the stethoscope, which has a rubber tubing connected to an end piece that is placed on the area that will be examined. The tubing conveys the sounds from the patient's diaphragm to the doctor's ears.

Loudness cannot be subjected to direct measurement, since it rest on the ear and the independent judgment of the listeners or receivers of sounds. Sound intensity is the rate at which the energy of sound waves is transmitted on through a unit per area perpendicular to the direction of wave motion. The bel named after Alexander Graham Bell, is the unit of sound intensity. Decibel is used when smaller unit of sound intensity.

Noise pollution is one of the problems which some communities are facing right now. Noise is defined as the irritating sound caused by irregular disturbances at the source. It commonly affects an individual's nervous system. It may even upsurge one's pulse rate and heart .Noise will be lessen if beat by decreasing the sound intensity. The ear has an inborn protective mechanism against loud sounds. However, there is a time delay between the onset of a loud sound and the ear's protective reaction. A very sudden loud sound such as explosion may damage the ears. The human ear can tolerate an intensity range of 0 to 120 decibels. Beyond 120 decibels, the ear feels pain and the eardrum might break. Music is naturally pleasant to the ears because it results from regular disturbances from the source. However, if music is very loud, it may become unpleasant.

Table of sound levels L (loudness) and corresponding sound pressure and sound intensity			
Sound Sources (Noise) Examples with distance	**Sound Pressure Level L_p dB SPL**	**Sound Pressure p N/m^2 = Pa sound field quantity**	**Sound Intensity I W/m^2 sound energy quantity**
Jet aircraft, 50 m away	[illegible]	200	100
Threshold of pain	[illegible]	63.2	10
Threshold of discomfort	120	20	1
Chainsaw, 1 m distance	110	6.3	0.1
Disco, 1 m from speaker	100	2	0.01
Diesel truck, 10 m away	90	0.63	0.001
Kerbside of busy road, 5 m	80	0.2	0.0001
Vacuum cleaner, distance 1 m	70	0.063	0.00001
Conversational speech, 1 m	60	0.02	0.000001
Average home	50	0.0063	0.0000001
Quiet library	40	0.002	0.00000001
Quiet bedroom at night	30	0.00063	0.000000001
Background in TV studio	20	0.0002	0.0000000001
Rustling leaves in the distance	10	0.000063	0.00000000001
Threshold of hearing	0	0.00002	0.000000000001

Figure 7.7: Intensity scale of sound (dB).

Sound Waves and its Speed

The speed of the sound, just like light, it can be measured utilizing the basic wave equation.

$$\mathbf{v} = \lambda f$$

where

λ is the wavelength

f is the frequency of the wave.

Wavelength is the distance between two resembling points on a wave train. Frequency pertains to the number of waves transmitting in a given point at a given time. A sound wave is a pressure disturbance that transmits through a medium with a manner of interaction through particle-to-particle interaction. As one of the particles becomes strained, it exerts a force on the next particle. Speed of a sound pertains to how fast the vibration has passed from particle to particle. However, frequency pertains to a number of vibration that each particle makes per unit of time, makes per unit of time, speed pertains to the distance in its given unit that the vibration transmit per unit of time.

Since the speed of wave is defined as the distance that point on a wave travels at unit of time. The unit meters per second is often used to express speed of the sound wave.

The equation follows below:

Speed = distance/time

The faster the sound moves, the more distance it will cover in the same period of time. If the wave was observed to travel a distance of 700 meters in 2 seconds, the speed of the wave would be 350 m/s, a slower wave could cover less distance- perhaps 660 m, in the same time period of 2 seconds and this has a speed of 330 m/s.

The pliability and density of a medium influenced sound generation. Therefore, the speed of sound varies in different media. Greater temperature also upsurge the speed of sound. For solids and liquids, the change in speed is small and can commonly be neglected; for gases, the speed increases with the temperature because the change is large. A number of experiments were conducted, the results show that the speed of sound in air at 0°C is approximately 330 m/s. When the temperature rise in every degree above 0°C, the degree of sound in air increases by 0.6 m/s. Therefore, at any temperature T in degree Celsius, the speed of sound in air is conveyed by the equation.

$$v = 330\ \text{m/s} + [(0.6\text{m/s})/0^{\circ}\text{C}]\ (T)$$

For instance, what is the speed of sound in air at 30°C? Using the equation above and substituting the value of T=30°C, we have

$$v = 330\ \text{m/s} + (0.6\ \text{m/s}\ 0^{\circ}\text{C})\ (30^{\circ}\text{C})$$

$$= 330\ \text{m/s} + 18\ \text{m/s}$$

$$v = 348\ \text{m/s}$$

Table 7.1: The sound wave's speed varies rendering to the media through which they transmit

Medium	Speed of Sound (m/s)
Air	330
Seawater	1 540
Freshwater	1 410
Brass	3 650
Copper	4 000

Wood	4 650
Steel	4 950
Iron	5 000
Aluminium	5 140
Glass (soda)	5 150

HOW SOUND WAVES BEHAVE

Interference

Interference is a phenomenon that happens when two waves meet each other while travelling on the same medium. The interference of the waves makes the medium to take on a shape that will be the outcome from the net effect of two individual waves on the medium's particles. When two or more different sources are present at the same time, they interact with each other to create a new wave. The new wave is the sum of all the different waves. Interference is also called as wave interaction. If the compressions and rarefactions of the wave tie up, they strengthened each other and create a wave with higher intensity. This type of interference is known as constructive interference.

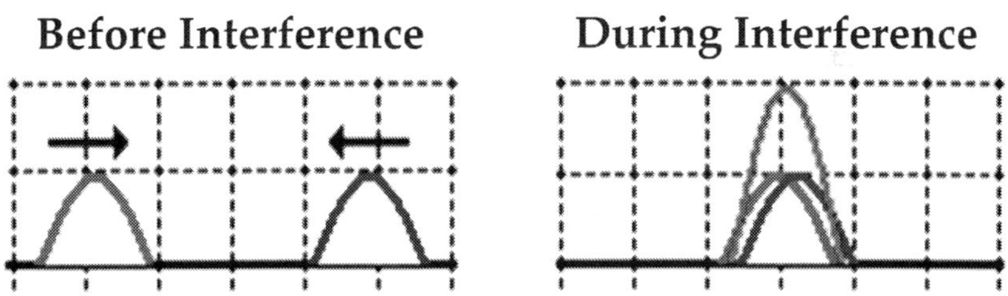

Figure 7. 8: Shows the interference of the sound waves before and during the interference.

Two Types of Interference

1. Constructive Interference

Constructive interference is a type of interference that happens at any location upon the medium where two intrusive waves have a displacement in the same direction. Due to this, both waves have an upward displacement; subsequently the medium has an upward displacement that is greater than the displacement of the two intrusive pulses. Constructive interference is observed at any location where the two intrusive waves are moved upward. However, it is also perceived that when both intrusive waves are moved downward.

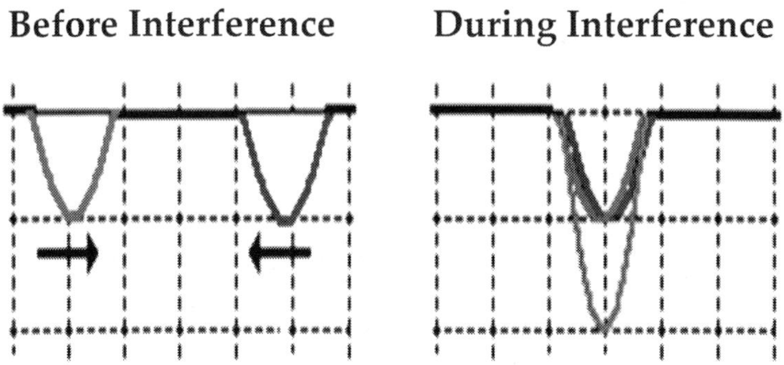

Figure 7.9: shows the Constructive interference.

2. Destructive Interference

It is a type of interference that occurs at any location along the medium where the two intrusive waves have a displacement in the opposite direction. For example, when a sine pulse with a maximum displacement of +1 unit meets a sine pulse with a maximum displacement of -1 unit, destructive interference happens.

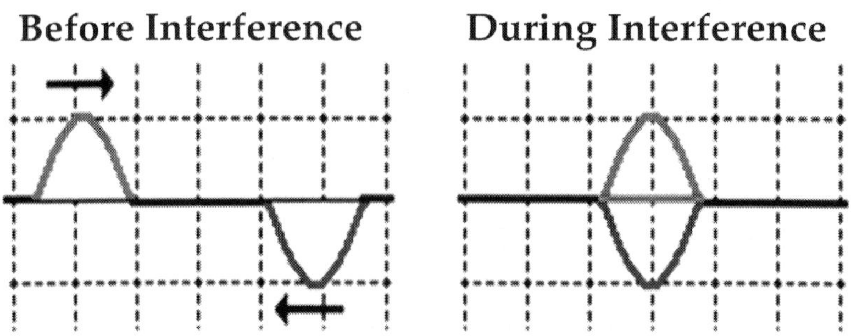

Figure 7.10: shows Destructive interference.

Figure 7.10 shows the intrusive pulses have the same maximum displacement but in opposite directions. The outcome is that two pulses totally destroy each other when they are totally overlapped. At the moment of overlap, there is no resulting displacement of the particles of the medium. The destruction is not a perpetual condition. Actually, to say that the two waves destroy each other can be somewhat deceptive. When it is said that the two pulses destroy each other, it is meant that when the two waves overlapped the effect of one of the pulses on the displacement of a given particle of the medium is destroyed or cancelled by the other pulse.

Resonance

Musical instruments are agreed into vibrational movement at their normal frequency when a player hits, strikes, strums, plucks or somehow distracts the object. One of the many standing wave patterns is related to each natural frequency of the object. One of the many standing wave patterns is related to each natural frequency of the object. Harmonics of the instruments are the natural frequencies of musical instruments. An instrument can be required into vibrating at one of its harmonics if another interconnected object pushes it with one of those frequencies. Resonance happens when one object pulsating at the same natural frequency of a second object required that second object into vibrational motion. Resonance comes from Latin word and it simply means to "resound"- to sound out together with a loud sound. The common means of propagation of sound in musical instruments is called resonance.

SHOCK WAVES AND DOPPLER EFFECT

Doppler Effect

Doppler Effect is an occurrence perceived whenever the source is travelling with respect to an observer. The Doppler Effect can be characterized as the effect generated by a travelling source of waves in which an outward shift in frequency is present for the observer and the source is diminishing. The Doppler Effect can be perceived to occur with all the types of waves mostly in water waves, sound waves and light waves.

Doppler Effect is very familiar to all of us because of our experiences with sound waves. Try to recall an instance where in an emergency vehicle was moving towards you on the high way. As the vehicle draw near with its siren blasting, the pitch of the sound was high and then as it passed by, the pitch of the siren sound was low. That describes the Doppler Effect, it is a shift in the outward frequency for a sound wave generated by a travelling source.

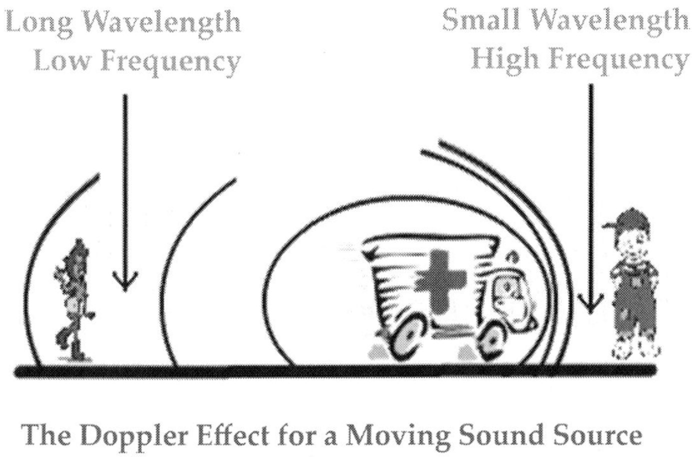

Figure7.11: shows the Doppler Effect of the moving sound waves.

Shock Waves

Doppler Effect is perceived whenever the speed of sound of the source is moving slower than the speed of waves. However, if the source actually moves at the same speed as or faster than the waves itself can move, a different occurrence is observed. If a travelling source of sounds moves at the same speed as sound then the source will always be at the leading edge of the waves that it generates. For an instance, the moving aircraft that is travelling at the same speed as sound. The circular lines in the figure represent compressional wave fronts of the sound waves. Observe that the circles are clustered up at the front of the aircraft, this phenomenon is called shock waves. Shock waves are also generated if the aircraft moves faster than the speed of sound. If a moving source will always be ahead of the waves it creates.

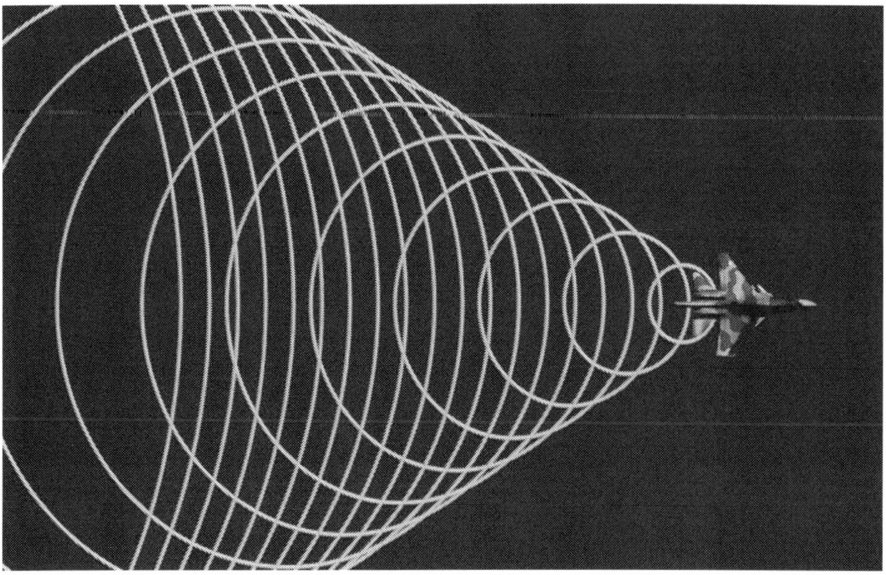

Figure 7.12: shows the phenomenon called Shock wave.

Sound Waves Boundary Behavior

When a sound wave transmits through a medium, it will often reach the end of the medium, it will often reach the end of the medium and come across an obstacle or probably another medium through which it could transmit. As one medium, ends another medium starts; the interface of the two media is called as boundary and behavior of a wave at that boundary is described as its boundary behavior. There are four possible boundary behavior by a sound wave could behave, reflection which is defined as the bouncing off the boundary, diffraction means the binding around the obstacle without crossing of the boundary, transmission which is characterized as the crossing of the boundary into the new material or obstacle and refraction which is defined as occurrence with transmission and is regarded by the succeeding change in speed and direction.

There are two types of reflection for waves that could travel on ropes: fixed end reflection and the free end reflection. A pulse transmitting through a rope will finally reach its end. When the pulse reach the end of the medium, two things are likely to happen:

A part of the energy carried by the pulse is reflected and returns towards the left end of the rope. The disturbance which returns to the left is called as the reflected pulse. A part of the energy carried by the pulse is transformed into the new medium. If the rope connected to a pole, the pole will receive some of the energy and begin to vibrate. If the rope is not connected but rather connected by loose ends then a part of the energy is transmitted into the air and that is the new medium, making a slight vibrations of the air particles.

Fixed End Reflection

Fixed end reflection is best demonstrated by a rope tied to a rigid object. When the disturbance reached the other end of the rope, which is tied to the rigid object, the energy on the last particle of the rope will be unable to move the vibration back because it is attached to a fixed object. This end of the rope is referred to as a fixed end reflection is observed in the other end of the rope which wave cannot travel back from its initial position. A pulse called incident pulse is presented at the rope's left end and traveled through the rope towards the medium's right end; this how waves behave in fixed end reflection. Incident pulse is created because of the occurrence of the boundary with the rigid object. Two things happened when incident pulse reached its boundary.

A. Reflected pulse is then created when a reflected part of the energy being carried by the pulse returns to the rope's left end. It is a wave or vibrations that transmitted back to the left end of the rope after rebounding back from the rigid object that the rope was tied into.

B. Vibration of the pole as the pulse transported the part of the energy it carried to the rigid object where the rope is tied into.

Free End Reflection

Free end reflection can be shown when a cord is affixed to a rigid object but this time instead of tied securely, ring will be used. This ring will make the rope moves loosely that is why when the remaining particle of the medium reached the end of the rope it will be able to move the object because it is attached to the object freely. And it refers to free end.

SOUND'S REFLECTION, REFRACTION AND DIFFRACTION

As the sound wave reaches the end of the medium or when it meets a hindrance in its path, it does not stop. Relatively, a sound wave will undertake specific behaviors when it meets and end or experiences a hindrance. The possible behaviors that the sound wave may undergo include reflection of the obstacle, diffraction around the obstacle and refraction which is accompanied by transmission into the obstacle or new medium.

When a wave has reached its boundary between one medium to another medium, a part of the wave undergoes reflection and a portion of the wave undergoes transmission across the boundary.

Echoes demonstrate to be troublesome in large halls and assembly rooms. Sound can be reflected repeatedly. The reflected sound perseveres until it disappeared. The repeated sound reflection is called reverberation. Reverberation typically occurs when the walls of the rooms are good reflectors of sound. However, a certain degree of reverberation is also essential. Assembly room and auditory may appear empty without it. The range of 1 to 2 seconds is the recommended time for reverberation in auditoriums. This is the time required for a repeated or reflected sound to fade away and diminish. Too much reverberation is also not recommended because it is undesirable. It points out poor acoustics. In industrial fields, the reverberation of machinery can cause auditory fatigue and worker's efficiency will diminish. One way to lessen reverberation is to cover the walls with sound-absorbent materials. Porous materials such as cork, fabrics and felt absorb sounds best. They transform some energy into heat. Therefore energy transport is reduced.

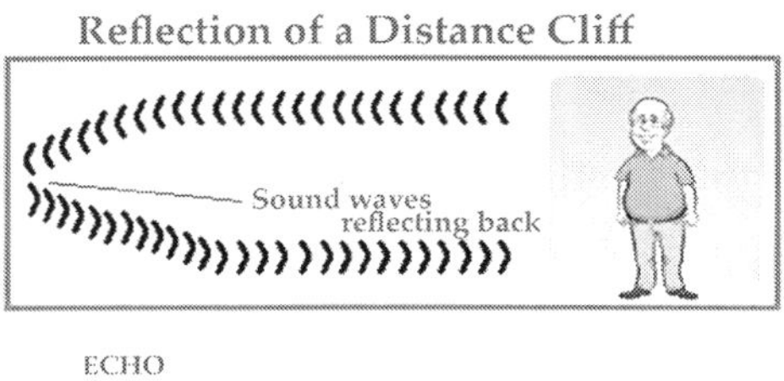

Figure 7.13: shows how reverberation of sound.

Diffraction

So what could be the proof that the sound will be spread out of when it passed through the gap? How do they bend around an obstacle? Well the evidence is that when you hear the neighbor's radio even there are houses that separate your house from the source of the sound is that sound waves can bend around obstacles or get diffracted. However, the sound that can be heard is not as loud it plays on their house. Like water waves, when sound waves pass through gaps become weaker compared to the original sound. The farther the sound transport, the weaker they become.

Refraction

Refraction of sound waves includes the change in path of waves as they travel from one medium to another. Refraction or bending of the path of the waves, is accompanied by a change in speed and wavelength of the waves. Therefore if the medium are changed, the speed of the waves are changed. Refraction of sound waves is best observed in situations in which the sound wave travels through a medium with regularly changing properties. For an instance, sound waves are known to refract when moving over water. Though the sound wave is not exactly changing media, when it is transferring from a medium with unstable properties, therefore sound wave will experience refraction and change its direction. Meanwhile, water has a controlling effect upon the temperature of air, the air directly above the water tends to be cooler than the air far above the water. Sound waves moves slower in a cooler air than they do in warmer air. For this reason, the part of the wave front directly above the water is slowed down, while the part of the wave fronts far above the water is speeds ahead. Consequently, the path of the wave changes, refracting towards the water.

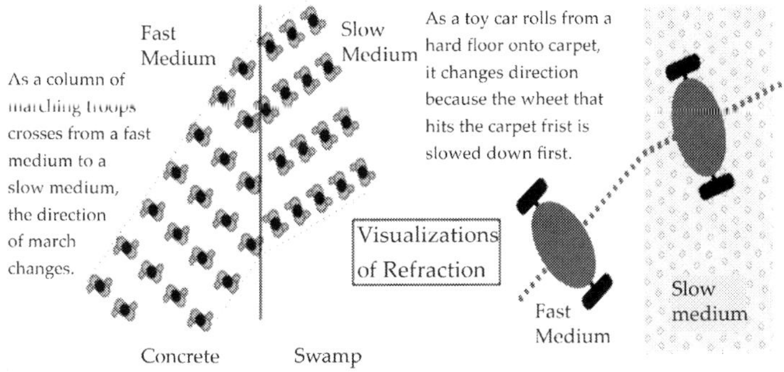

The illustration above will help in comprehension of the nature of refraction. A pillar of troops arriving a medium where their speed is slower as shown will turn toward the right because the right side of the pillar hits the slow medium first and therefore they slowed down.

CHAPTER 8

APPLIED SCIENCE

INTRODUCTION

Scientists in different places in the world work for one goal, that is for innovation that will make the life of man a lot easier. Different works deal with the things with different applications that will aid for the betterment of society. Appreciation of the sacrifices and hard work that these scientists endured during discovery is the only thing that modern world can do for them. Spread their efforts and use their inventions for the common good of the people.

In natural world, basic science is utilize to harness information to describe the occurrence in the natural environment. This statistics is then apply to use for practical activities through applied science. Applied science is commonly engineering, which harness technology, however there might be a debate about the issue between basic science and applied science which includes research and development.

Medical sciences like medical microbiology are examples of applied sciences. These sciences utilize concepts of biology toward medical information and inventions, even though the application is not essentially medical technology, which is advanced more particularly through biomedicine or biomedical engineering. For an instance, epidemiology is an application of the formal sciences of statistics and probability theory. Epidemiology deals with the study of the patterns, causes and effects of health and disease conditions in defined populations. Another example is the application of both biological and statistical methods in an applied science called Genetic epidemiology.

DEFINITION OF APPLIED SCIENCE

Applied science is specialty of science that utilize occurring scientific knowledge to harness more practical applications such as technology or innovations that lead to invention. It is derived from the studies of natural science, area that are basic science and also known as pure science. They are harnessed statistics that foresee and probably somehow explain the phenomena in the natural world. Science specifically applied science applies different field of sciences to real world practice.

Applied science includes wide range of related science such as engineering, medicine, early childhood education, statistics and probability theory and genetic epidemiology.

The application of scientific knowledge transferred into physical environment is pertaining to applied science. Testing a theoretical model through the utilization of formal science or solving a natural problem through the application of natural science.

Applied science is far different from basic science which search applies more to describe basic objects and forces without the applying practical applications to deal such problem. The aim for knowledge is to be just known. Applied science can be characterized as biological science or physical science. Both sciences are very essential in lives of people living on Earth. It is evidently known that human being depend on abiotic component of the ecosystem, which is also pertained to physical science, in order to live.

BRANCHES OF APPLIED SCIENCE

Engineering deals with the application of mathematical theories and scientific knowledge, social and hands-on knowledge which aims to create innovations such as new inventions, design, build , maintain, studies and to develop structures, machines, tools, system, components, process, solutions and organizations. The concept and area of engineering is enormously broad and comprises a range of various and more specialized engineering's field of study, each consists of more specific emphasis on specific area of applied science in association with technology and the type of application that adheres to solve or give solution to a current situation or problem.

Engineering can be defined as an act of devising something with cleverness as it is derived from the Latin word ingenium which means "cleverness" and ingeniare which denotes "to contrive or devise".

Figure 8.1: It signifies the importance of engineering in modern history.

Engineering is also defined as the ingenious application of scientific ideologies to plan or develop structures, machines, equipment or apparatus, manufacturing processes which can be utilized as one or combination. The awareness of their design was very essential in recognizing their construction or operation. To foretold the behavior of machines which is applied to solve such situation under the operating conditions. And most of all give due respect to the aimed functions, to the operation in economics or protection of life and property.

The concept of engineering has been present since early period as humans basically devised inventions such as simple machines which include wedge, lever, wheel and pulley. They are devised mainly to make the work of man a lot easier and to save effort, energy and time. These inventions is evidently consistent in connection with modern description of engineering. Another etymological definition of the word engineer is that it came from a word which is identified since 1390, the word engineer literally means someone who operates an engine. The said word was initially pertained to "a constructor of military engines", however this meaning of engineer is now obsolete though the U.S. Army Corps of Engineers has survived and still present to modern days.

As time passed by, the design of establishments, bridges and landmarks evolved into more matured technical discipline, therefore the words civil engineering was added to the vocabulary in order to differentiate between those fields of specializing in the world of construction of projects both non-military and military engineering.

Figure 8.2: Ancient era, the ancient Roman built aqueducts to distribute stable supply of fresh water to cities and towns in the empire.

Figure 8.3: Another ancient structure in ancient era, the Great Wall of China.

Figure 8.4: Modern era, the Falkirk wheel in Scotland.

Engineering is often perceived to be consisting of four main branches. These are the following:

1. Chemical engineering

Deals with the application of physics, chemistry, biology and engineering principles in order to perform chemical processes on a marketable scale like petroleum, micro fabrication, refining, fermentation and production of biomolecule.

2. Civil engineering

Generally pertains to the design and construction of public and private works which include infrastructure like roads, bridges, railways, airports, bridges, dam and establishments.

3. Electrical engineering

Deals with the design, study and manufacture of different electrical and electronics systems like electrical circuits, generators, motors, devices connected with electromagnetic waves, optical fibers, computer systems, telecommunications and instrumentation.

4. Mechanical engineering

Work connected in the design and manufacture of physical or mechanical systems such as power and energy systems, aircraft products, weapon system, transportation, engines, and power trains, vacuum technology, manufacturing and mechatronics equipment.

APPLIED MATHEMATICS

Applied mathematics is one of the branches of applied science and also a branch of mathematics that contracts with mathematical methods that discover a potent use in science, engineering, business and industry. Mathematical science and specialized knowledge comprise applied mathematics. It characterizes the works of mathematicians professionally designing and studying mathematical models to be applied on practical problems.

Just like the concept of applied science, practical development on applications of mathematical theories was founded through basic or pure mathematics. The researches done in connection with pure mathematics was developed in activity associated with applied mathematics directly because it tackles about the innovation that can be applied to a particular problem in the field of mathematics.

Based on its history, applied mathematics is composed different fields of mathematics which are associated directly to the progress of Newtonian physics. Before the mid-19th century, mathematicians were not distinguished from physicists. This situation left an education legacy in the United States where subjects like classical mechanics were tackled under applied mathematics departments rather than in physics department at universities. Fluid mechanics still be tackled under applied mathematics departments. Mathematical finance and quantitative finance are taught under mathematics department in the universities. Computer science and engineering are originally utilizes concepts of applied mathematics.

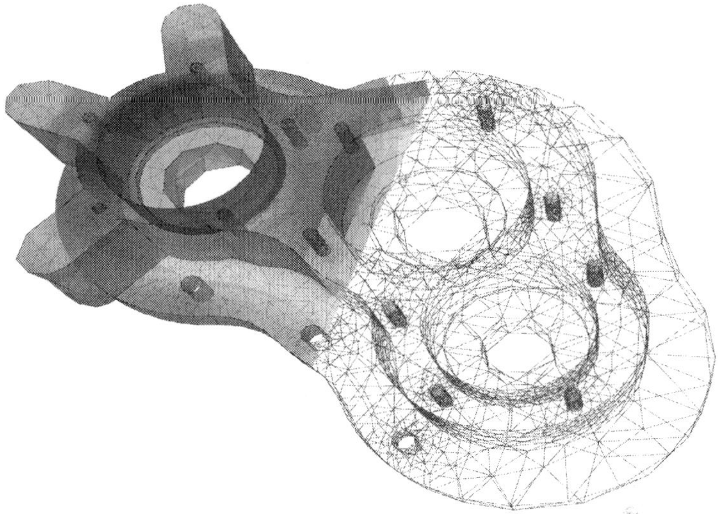

Figure 8.5: A numerical solution to the heat equation on a pump causing model using the finite element.

APPLIED PHYSICS

Applies physics is a science which is projected for a particular technological and practical use. It serves as a bridge or a link between physics and engineering. The concept applied can be differentiated from pure through a simple combination of factors like motivation and attitude of researchers and the kind of relationship to the technology or science that can affect the nature of work. The origin of applied physics is from the fundamental truths and simple concepts of the physical sciences but is apprehensive with the usage of principles in science concerning with practical devices and systems, and in the applied concepts of physics in other fields of science.

The difference of engineering from applied science is that physicist may not be manipulative in creating something specifically but somewhat using physics or conducting research linked in physics with a purpose of developing new innovations or solving an engineering problem. Applied mathematics has the same approach with applied physics. In other terms, applied physics is originated from basic truths and concepts of the physical sciences but is apprehensive with the usage of these scientific ideologies in practical devices and systems.

MEDICINE

Medicine is the science and practice of the diagnosis, treatment, and prevention of disease. It is originated from the Latin word medicus, which means "a physician". Medicine embraces various practices under health care that developed to sustain and establish health by the hindrance and cure of illness. Modern-day medicine appertain biomedical science and research, genetics and medical technology to identify, treat and prevent injury and disease through typical pharmaceuticals or surgery but also through therapies as varied as psychotherapy, external splints and traction, medical devices, biologics and ionizing radiation among others.

Medicine has occurred for thousands of years, which was perceived as an art which is pertaining to area of skill and knowledge usually having association to the religious and philosophical beliefs of local culture. For an instance, a medicine man would put herbs and tell prayers for healing, o an ancient theorists and physicians would apply bloodletting conferring to the theories about humorous concepts. Medicine, in recent centuries of the arrival of modern age medicine, has perceived to be the combination of art and science which is comprised both basic and applied science under the authority of medical science. To show that medicine is both an art and applied science let us analyse the situation, stitching sutures is an art learned through practice and the knowledge of the result when the tissue at cellular and molecular level was stitched ascends through science.

Traditional and folk medicine are the prescientific form of medicine. The use of these medicine continue to be used rather than scientific medicine and therefore are called alternative medicine. For an instance, the proof on the efficiency of acupuncture has different and unpredictable depending on the condition of the customer, however mostly safe if professional practitioner will conduct it. In distinction, treatments unconfined with the boundary of being safe and efficacy are called quackery.

COMPUTER SCIENCE

Computer science deals with the study of theory, experimentation, and engineering that comprise the foundation for the design and the use of computers. Computation was based on scientific and practical approach which applications and the step by step study of the feasibility, structure, expression, and programming of the procedures guided by methods that trigger the acquisition, representation, processing, storage, communication of and access to information. Computer science is defined logically as the study of automating algorithmic process that scale. The art of computational systems and theory of computation was the specialization of computer scientist. The field of computer science is divided into a different theoretical and practical disciplines. Computational complexity theory, one of the fields of computer science which explores the basic properties of computational and problems that intractable, is highly intangible or abstract while applications of field such as computer graphics give emphasis on real-world visual presentations. Other fields still concentrate on challenges in employing computation. Cited examples are programming language theory contemplates different approaches to the characterization of computation, meanwhile computer programming examines different aspects of the usage of programming language and complex system. Interaction between human and computer reflects the challenges in making computers and computations valuable, functioning and entirely accessible to people.

History of Computer Science

The ancient fundamentals of computer science precede the discovery of the modern digital computer. Apparatuses for manipulating or calculating fixed numerical tasks like abacus have occurred since ancient times, helping in computations such as multiplication and division. Moreover, algorithms for execution of computations have been in existent since time immemorial, even before the progress of refined computing device. Wilhelm Shickard planned and created the foremost functional mechanical calculator in 1623. Then in 1673, Stepped Reckoner a digital mechanical calculator was demonstrated by Gottfried Leibniz. Due to considerable reasons, Leibniz was considered the first computer scientist and information theorist, he also documented the binary number system. Thomas de Colmar in 1820, has established the mechanical calculator industry when he launched his abridged arithmometer, which was the first calculating instrument durable enough and reliable enough to be utilized daily in an office environment. The first automatic mechanical calculator was designed by Charles Babbage in 1822. His Difference

Engine which finally gave him the concept of the first programmable mechanical calculator. The development of machine was initially started in 1834 and after two years, the sketch of several prominent features of the modern computer came out. The most difficult part was the adoption of punched card system that imported from the Jacquard loom, making the computer substantially programmable. Ada Lovelace, in1843 during the transformation of a French article on the Analytical Engine, wrote a note from her many written notes, she included that an algorithm to compute the Bernoulli numbers, these are considered to be the foremost computer program.

Then in around 1885, Herman Hollerith conceived the tabulator, which is utilized punched cards to treat statistical information; finally his company became a part of IBM. Howard Aiken, in 1937 and one hundred years after Babbage's incredible dream, convinced IBM, which was responsible in making all kinds of punched card apparatus and was also in the industry of calculator to harness his enormous programmable calculator, the ASCC/Harvard Mark I which is based on Babbage's Analytical Engine.

Figure 8.6: Ada Lovelace was credited from her first writing about algorithms intended for processing of computer.

Babbage's Analytical Engine uses cards and a central computing program. When the machine was completed, some commented on it as "Babbage's dream come true". As the new and more prevailing computing devices were progressively created, the term computer arrived to pertain to the apparatuses rather than their human precursors. When it was realized that computers could be utilized for more than just a mathematical calculations, the study of computation was broadened the field of computer science in general. The establishment of computer science has begun as a diverse academic department in the 1950s and early 1960s. Cambridge Diploma in Computer Science was the world's foremost computer science degree program that was offered at University of Cambridge Computer Laboratory in the year of 1953. The United States piloted their first computer science degree program at Purdue University. The availability of the practical computers made many applications of calculating have become distinctive fields of study in their own respective privileges.

Granting that critics primarily conceived that computers were impossible for them to be a scientific field of study, in the almost last step of fifties, computer science progressively accepted among the greater academic department. It is the most famous IBM brand that shaped portion of the computer science uprising until now. International Business Machines or IBM launched the IBM 704 and later the IBM 709 computers, which were used extensively during the exploration time of such devices. However some problems still occurred while using the computer, like when the manipulator of the computer misplaced one letter in the instruction it was frustrating that the program would crash and the explorer need to start the whole process all over again. This problem was treated as ordinary when the computer discipline was in the height of its developmental stages during the late 1950s. As years passed, generation has seen essential developments in the usage and efficiency of the computing technology. The shift in the users of computer technology has witnessed by the society, the shift was from experts and professionals to a common user base. At first, computers were quite expensive and some point that human aid was needed for effective use, especially in part of expert computer operators. As computer acceptance become more prevalent and reasonably priced, less human support was needed for ordinary usage.

Figure 8.7: Charles Babbage sometimes referred as the Father of Computing.

Figure 8.8: The IBM 704.

Figure 8.9: The IBM 709.

BENEFITS AND DANGERS OF TECHNOLOGY

Benefits of Technology

Education

Technology is of great benefit in the area of education and society in many ways. One of the many benefits is the application of internet. It is an enormous source of information that is available from everywhere on the entire planet. It appears to be boundless sources that can be utilized for good particularly when employed for resolutions of educational betterment. However the availability of internet for almost everywhere arrived with its fair portion of perils, and the awareness of how to manipulate the internet safely is an inevitably in today's world. But the positive effects compensate the negative effects.

The Internet offers people the chance to educate themselves, and they can do it even in the comfort of their own home if they wish to install it to their houses. It permits us to learn limitless about all the things that we want to learn. You can study your lessons in advance just search it online. If you are interested to learn a specific lesson, subject or course that the school or college does not cover, you have all the opportunity to study or teach yourself through your own online research or by signing up to an online course. It is your choice to study free of charge or with fee. Due to the helpful and cooperative characteristics of the majority of people who use the Internet, help is available over time. You can post queries in online forums or other websites and get responses instantaneously. Connectivity or communication is another evident benefit of the Internet. It permits interaction between teachers and students or other educational professionals even in other countries, they can be either in the vicinity or geographically far. Therefore, Internet is concluded to be beneficial in developing the whole education system a lot faster and more accessible, which is the most important and major benefit Internet can offer.

Interactive boards and the availability of computers to students inside the classroom allow a modern way of teaching that is much more collaborative compared to traditional methods. The assimilation of tablet PCs into the classroom adds further opportunities for interactive learning.

This permits for more efficient learning as the students become engaged in the modern learning process instead of conventionally just battling their selves to the outside factors that create hindrances on focusing to listen to their teacher.

Mainly through the use of Internet, there are many personalised learning techniques have arisen. It is the result of technology in the field of education. The foundation for this is that everyone can learn at their level of understanding at their own place and time without competing and get compared to others. The purpose of teacher-made learning process through the usage of computer and Internet is to encourage students to be more accountable for their own learning, to be more independent and to provide students to learn in ways that ensemble their individual learning styles. Even though technology that aids education is still on its comparatively early stage, it is evidently having an enormous impact on the educational system. The proof of its positive impact is that how seemingly education is starting to depend on the Internet. This emphasizes the power of technology and that technology will be incorporated to all fields of education or in wherever it is beneficial.

Benefits of Technology in Business

Technology is a wide-ranging concept that has arrived to denote to innovations in science that permit for an improved or programmed solution. The most evident benefit of technology to small business is the improved productivity, which transforms into lower cost structure, aside from this there are other benefits of technology that are helpful in the bottom line as well. Improved speed, the comfort of sharing and storing important details and decrease human error through programmed systems that add up to the reduction of costs and increase in profit is also notable.

Speed and Time

Small industries or businesses contend with huge businesses by making themselves more instantaneous and responsive. Moreover, a small business can embrace change more quickly compared to huge businesses. From ancient times, when electronic mail is not existing yet, postal service is the only way to communicate with someone else in different geographical location. The farther the location of the client or supplier, the lengthier the communication process. Today with the initiation of technology to treat information, it is shared at an astonishing degree. This method saves time, proposing the ability to create decisions in a faster way.

Convenient Storage

The file of extremely bulky amounts of paperwork and the method of double or triple entry systems were eliminated by technology. Contracts and costumer's profile can be stored in virtual data bank and can be accessed whenever there is a need to update the information, therefore the need to purchase or rent storage space was no longer an issue to the owner of the business.

Improved Sharing of Information

Information become available whenever and wherever you are through the aid of technology. These information whether written or broadcast can be shared faster and with minimal resources. Consummation of marketing strategy can be conceived through posting an ad that can reach millions of ready buyers or subscribers on the Internet or through social networking sites such as Instagram, Twitter and Facebook. E-learning and other practices of online training have redesigned the availability of the average small business workforce as employees can heed to classroom lectures and share ideas and opinions with classmates from the beneficial comfort of their home or office. This erases the need for small business to hire training employees.

Programmed systems

Automated program on certain functions was deliberately conducted by technology, whereas the eliminating the ancient practice of hiring someone to do some paper works. For example, bookkeeping purposes can now be handled by software applications like Quicken and Quickbooks. Sale Force which is a contact management site, was utilized to program the sales function of the business. This provides the owner of the small business the ability to concentrate on strategy and minimize the labor expenditures.

Easy Access to Information

Getting access to relevant data at any time anywhere. This can be experienced because of the technologies like broadband internet. Lots of information is being published and recorded online, sites such as Wikipedia and You tube have large original concept which can be utilized in research studies or entertainment. Statistics is power and those who see information and utilize it well always prosper. With smart modern gadgets such as iPad, iPhone, galaxy tablet, subscribers can easily have access to information because they utilize the internet. Meanwhile, as a subscriber on a train can easily read breaking news while travelling, they can also purchase and sell stocks while in the bedroom using internet. These smart devices make it easy to access the internet and this abridges the way we get statistics.

1. Emboldens innovation and creativity

Since technology is challenging, it sparks the brain to work to its full potential. In the past, it used to be very difficult to start a business, one had to have lots of capital and they even had limited access to business information. Today, it is very easy to start a business while at home. Let's look at companies like Etsy.com which enable creative people to sell their works online, this encourages creativity. Another good example is kickstarter.com which helps creative people get funds for their projects through crowd funding. On this platform, creative developers post projects seeking funding from the community, this helps them generate lots of cash for their good ideas which latter leads to the creation of new Jobs. The other creative works which have been facilitated by modern technology include Google, Apple, Facebook, Microsoft, Amazon, and etcetera.

2. Enhanced communication

Communication is like water to life. We cannot develop without communication. Modern technology has blessed us with advanced communication technology tools. These can include e-fax machines, electronic mail, mobile phones, video conferencing, instant text messaging applications, social networking applications. All these modern communication technology tools have simplified the way humans and businesses communicate. I can easily talk to my relative overseas using a mobile phone or video chatting services like Skype.

3. The accessibility of Travelling

Modern transportation technology makes it very easy to travel long distances. Transport is a very important both in our lives and in the business world. Transportation technology has evolved with years. In the past, it used to be slow and expensive to move long distances. Nowadays, I can cover a 10 miles distance within a few minutes or hours using electric trains or airplanes.

4. Enhanced housing and lifestyle

This is another great way how modern technology has simplified our lives. If you compare the type of housing we used to have in 1900 and the architecture of houses today, the difference is very big. New architectural technology has improved the types of housed we build nowadays. People with money can afford town floating housed and glass homes. Most of the items in our house are now automated, for example, doors use fingerprints which guarantee security. Remote webcams which you can use to monitor what goes on at your home.

5. Enhance Entertainment

Modern technology has played a big role in changing the entertainment industry. Home entertainment has improved with the invention of video games and advance music and visual systems like smart televisions which can connect live to the internet so that a user can share what they're watching with friends. Easy access and storage of music, services like iTunes allow users to purchase and download music on their iPods at a small cost, this is a win – win situation for both musicians and the users. Because musicians can easily sell their music via iTunes and the user can also have a wide selection of which music to buy without having troubles of going to a physical music store.

6. Efficacy and Productivity

Modern technology has helped businesses increase production. Humans are slow and sometimes they fail to deliver on time. So many businesses have integrated modern technology in their production line, most of the hard work has become so simple and the results are better than those of humans. Let's look at a farmer who uses modern technology right from the day of preparing the farmland to the day of harvesting. They save a lot of time and money during this process.

7. Expediency in Education

Learning is a process and it is part of our daily lives. Modern technology has made it simple for students to learn from anywhere through online education and mobile education. Also, students now use modern technology in classrooms to learn better. For example, students use iPads to share visual lessons and examples with peers in the classroom. This has made learning more convenient and fan. Also new modern educational technologies support individual learning which gives a chance to students to learn on their own with no need of tutors.

8. Social Networking

Modern technology has made it simple to discover our old friends and also discover new people to network with. This is a benefit to both individuals and businesses. Many businesses have embraced the social networking technology to interact with their customers. Users of social networks can share information with friends, live chat with them and interact in all sorts of ways.

9. Alter the health industry

Nowadays most hospitals have implemented modern technology in surgical rooms, this has reduced on mistakes made by doctors. Humans can easily make mistakes because of work overload and stress factors. Also, the business community has developed health applications which can enable us to monitor our health and weight. These applications can be used on mobile phones, so users can have them at any time of the day.

Four Disadvantages of Modern Technology

Loneliness is increased

Social Isolation is on the increase, people are spending more time playing video games, learning how to use new modern technologies, using social networks and they neglect their real life. Technology has replaced our old way of interacting. If a user can easily interact with 100 friends online, they will feel no need to going out to make real friends which at a later stage leads to loneliness. Modern technology has replaced many humans; robots are doing the jobs which used to be done by humans. Many packing firms have employed robots on production lines to increase production and efficiency, this is good news for businesses because it helps them make more money and serve customers in time, but it is bad news to employees because they get replaced by a robot.

Proficiency

Increased dependency on modern tools like calculators has reduced on our creativity. You can find a student when they can solve a very simple mathematical equation without using a calculator. This affects the way this student uses their brains and reduces the level of creativity.

World destruction weapons Modern technology has been the main aid in the increasing and endless wars. It aids the manufacturing of modern war weapons which will require testing. So when these weapons get into the hands of criminals, they will use them for their selfish reasons.

Impact of technology to the environment

Technology and human life cannot be separated; society has a cyclical co-dependence on technology. We use technology; depend on technology in our daily life and our needs and demands for technology keep on rising. Humans use technology to travel, to communicate, to learn, to do business and to live in comfort. However technology has also caused us concerns. Its poor application has resulted into the pollution of the environment and it has also cause serious threat to our lives and society. This calls for proper use of technology. The biggest challenge facing people is to determine the type of future we need to have and then create relevant technologies which will simplify the way we do things.

It is impossible to explore how each new advanced technology has impacted our lives and how it will impact the future. Technology impacts the environment, people and the society as a whole. The way we use technology determines if its impacts are positive to the society or negative.

For example, (**POSITIVE IMPACT**) we can use corn to make ethanol and this ethanol can be used as fuel. Fuel can be used to run machines and cars which will increase on output of manufacturing industries at a lower cost. (**NEGATIVE IMPACT**) However, if we decide to shift large quantities of corn to fuel production from food production, humans will be left with no food and this will cause world hunger which even be a worse situation.

Figure 8.10: Modern technology which is an aid for travelling via air.

Basing on the example above, technology by its self is not harmful to the society, but the way society uses technology to achieve specific goals is what results into negative impacts of technology on the society. Humans need to use energy to process products in factories, to run cars, to light homes and also run technological machines like computers, but the only way we can do this without affecting the environment and society, is by shifting from exhaustible energy sources to renewable and inexhaustible sources like Solar / Wind energy. Below I have listed both positive and negative impacts of technology on our society:

Positive Impacts of Technology on Society

1. Technology has automated agriculture.

Technology has mechanized agriculture : Modern agricultural technology allows a small number of people to grow vast quantities of food in a short period of time with less input which results into high yields and RIO "return on investment". Through government subsidies, small and medium sized farmers have managed to acquire ploughing, sowing, watering and harvesting machines. The use of technology in agriculture has also resulted into the manufacturing of genetic crops which can grow fast and they can be resistant to many pests and diseases. Also farmers have access to artificial fertilizers which add value to the soil and boost the growth of their crops and enable them to produce high quality yields. Farmers in dry areas have been in position to grow healthy crops, they use advanced water pumps and sprinklers which derive water from rivers to the farms, the all process can be automated to save time. A good example is Egypt, this is a desert country which receives little rain, but small and big farmers have used automated sprinklers to irrigate their farms. In Egypt, they grow a lot of rice, yet this crop needs sufficient water to grow well. The water is pumped from River Nile to the rice fields on a daily basis.

2. Technology has improved transportation

Technology has improved transportation: Transportation is one of the basic areas of technological activity. Both society and businesses have benefited from the new transpiration methods. Transportation provides mobility for people and goods. Transportation, like all other technologies can be viewed as a system. It is a series of parts that are interrelated. These parts all work together to meet a certain goal. Transportation uses vehicles, trains, airplanes, motorbikes, people, roads, energy, information, materials, finance and time. All these parts i have mentioned work together to move and relocate people and goods. Technology has helped in advancing all the four types of transportation and these include ; **(1)** road transport used by automobiles ,**(2)** air transport which is used by airplanes , **(3)**water transportation which is used by ships and speed boats and **(4)** space transportation used to go to the moon. The most used of all these is Road transportation, this one facilitates the movement of goods and people. Technologies like automobiles, buses and trucks have improved the way humans move and how they transport their goods from place to another. Also developing countries are getting funds from wealthy countries to improve their road transport which has resulted into development of rural remote areas.

3. Technology has improved communication

Technology has improved communication: Communication is used for a number of purposes. Both society and organizations depend on communication to transfer information. People use technology to communicate with each other. Electronic media like radios, televisions, internet, and social media have improved the way we exchange ideas which can develop our societies. In many countries, radios and televisions are used to voice the concerns of the society, they organize live forums where the community can contribute through mobile phones or text service systems like tweeter. During political elections, leaders use radio, television and internet media to reach the people they want to serve. A good example is the recent presidential election in USA, "President Barak Obama" has embraced the communication technology to reach and voice out his concerns towards the development of America. Communication technologies like Televisions, radios and internet can be used to persuade, entertain and inform the society. Small businesses have also used internet and mobile communication technology to grow and improve their customer service.

3. Technology has improved education and learning process

Technology has improved education and learning process: Education is the backbone of every economy. People need well and organized educational infrastructures so that they can learn how to interpret information. Many schools have started integrating educational technologies in their schools with a great aim of improving the way students learn. Technologies like smart whiteboards, computers, mobile phones, iPads, projectors and internet are being used in **classrooms** to boost student's moral to learn. Visual education is becoming more popular and it has proved to be the best method of learning in many subjects like mathematics, physics, biology, geography, economics and much more. The business community has invested money in various educational technologies which can be used by both teachers and their students. For example, on iTunes, you will find many educational applications which can allow students and teachers exchange academic information at any time, this has made learning mobile. Also programs like Long distance learning have opened boundaries to so many scholars around round the world.

Negative impacts of Technology on Society

1. Extinction of resources

The more demand for new technologies and advancement of current technologies, the more pressure we put on earth's natural resources. Look at the total number of mobile phones and computers being manufactured today, our population is increasing every day and all these billion consumers demand either a mobile phone or a computer in their homes or offices. This is good news to the manufactures, like Apple or Samsung, the demand for their gadgets is high, but to sustain this demand, they have to exploit Mother Nature for resources like aluminium, once these resources are extracted from the earth plates, they will never return back because it took a billion years to mature. That means that at one time, we shall be left with no natural resource which can be a problem to the future generation and economy. Likewise, the intensive farming practices will deplete the soil. This makes heavy applications of commercial fertilizers necessary to yield healthy harvests, but also these fertilizers have chemicals which are dangerous to the soil and human lives.

Technology has helped us live longer by improving health facilities and aiding in the research for solutions for most health problems which affect humans. This is good news for developed countries, but is bad news for developing countries which have not been in position to access these health care benefits brought by technology. In developed countries population growth is controlled by advanced birth control methods, this has helped them balance their population in relation to natural resources and other opportunities which come with a planned population. This is different in developing countries, the rate at which people produce is very high , the mortality rate is high , food is scarce and health care is poor.

2. Increase in population

Pollution affects the land we grow crops on, the water we drink and the air we breathe. The increased demand for new technologies and advancement of technologies has resulted into many manufacturing and processing factories. As they work so hard to create the best technologies for both society and business, they release harmful chemicals and gasses which have polluted our environment and this has resulted into climate changes (global warming). So the more technology we enjoy, the more we harm our environment. Experts have tried to implement ways of reducing this impact by encouraging factories to go green , to a small extent, this has been achieved through the development of green technologies like; green cars , green computers, but a great effort is still needed to reduce the pollution of the air and the earth.

WAVES, TECHNOLOGY AND COURSES

Aeronautical Engineering Degrees

Architecture Degrees

Chemical Engineering Degrees

Civil Engineering Degrees

Computer Science Degrees

Electrical Engineering Degrees

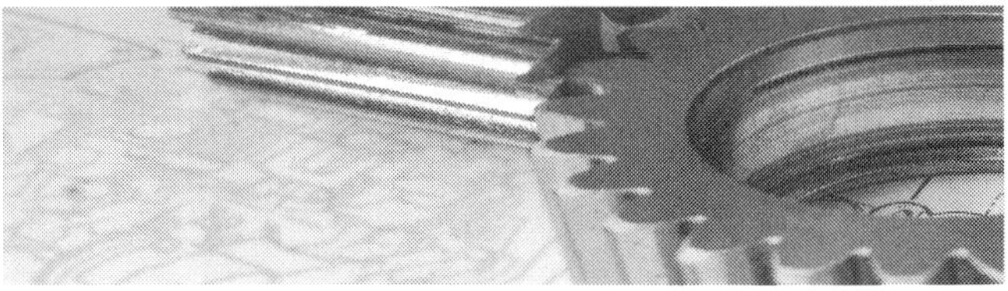

Mechanical Engineering Degrees

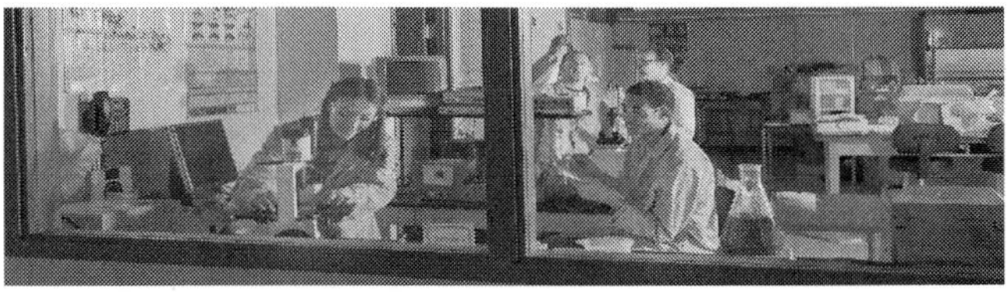

Biomedical Engineering Degrees

Environmental Engineering Degrees

Postgraduate Studies

Civil Engineering Degrees: Graduate

Masters in Computer Science

Engineering Management Degrees

Masters in Engineering

Masters in Mechanical Engineering

Masters in Electrical Engineering

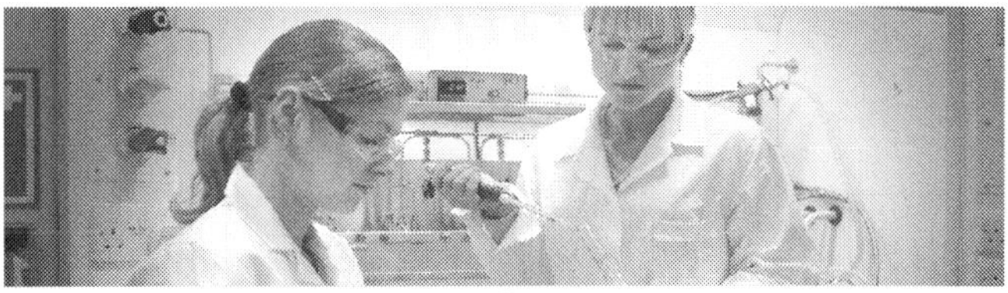

Masters in Chemical Engineering

Masters in Mining Engineering

(Source: www.google.com.)

Masters in Aeronautical Engineering

Masters in Manufacturing Engineering

Masters in Architectur

CHAPTER 9

ENERGY WAVES AND COMMUNICATION TECHNOLOGIES

INTRODUCTION

The unit centers on the use of energy in communication focuses on the utilization of sound and electromagnetic waves in telecommunications. The transformation of sound and energy carried by electromagnetic waves to electrical signs and back, as well as the revolution brought about by the progress in solid state electronics are described using the modern telecommunication technologies.

Communication is the transfer and receiving of meaningful and informative details on everything that is essential on this earth. People communicate through their sense, animals too use their senses in communicating. Through the idea of light and sound, they usually communicate with each other within the range of sight and hearing by visual and vocal means.

In attempt to find ways to communicate with the people who live far away many inventions had been proliferate in the field of technology and communication. Through different technological devices such as telephone, telegraph, radio, television, radar radio telescope, orbiting satellites, microwave relays and lasers as well as the advances made in the field of electronics, human communication is the recipient of immense favors. Unlike before the hindrance in time and distance barred efficient and faster interaction. Today the benefit of communication technology allowed people from the opposite poles to talk like they are just sitting side by side. The conversion of light and sound into electrical energy, and the utilization of electromagnetic waves for the transmission and reception of information is the foundation of these methods of communication.

Many scientists, engineers and inventors contributed to the development of telecommunications. They have spent hours of working in order to device some tools and equipment that may help people in communicating and send information even in the farthest place in the world. The immense effort and time was not wasted as the communication has reach its peak. No one even the people of yesterday cannot imagine the progress that the world is enjoying right now in the field of communication.

People of today enjoy the benefits and enjoyment in using social media cites in communicating to each other. You may even see through the monitors the people you love even though they are far away from you. These social media includes Facebook, twitter, Instagram, and other sites available for the human use.

THE CREATION OF TELEPHONE

Contributors to the Development of Telephone

1. Alexander Graham Bell

He is given the credit as the inventor of the telephone. The first telephone model was made by his assistant, Thomas Watson. Bell was a professor of vocal physiology in Boston University. He did experiments on the telegraph earlier developed by Samuel Morse. His little knowledge of electricity and his expertise in acoustics helped him succeed in building his liquid telephone model. The model is consisted of a wooden stand, funnel, a cup of acid, and some copper wire.

Stories have it that while working in the laboratory, some acid spilled on Bell and he called Mr. Watson. "Mr. Watson, come here, I want you"-

the first complete sentence transmitted by phone on March 10, 1876. Since then, the telephone instrument has evolved into different models using the principles of electricity and magnetism or electromagnetism.

Figure 9.1: Alexander Graham Bell.

2. Hans Christian Oersted

In 1820, Hans Christian Oersted discovered electromagnetism. While preparing for a lecture in his laboratory, Oersted unintentionally placed a magnetic compass under a current carrying wire. To his surprise, the magnetic needle deflected as if it were in the vicinity of a permanent magnet. Oersted concluded that current-carrying conductors produce a magnetic field. This is the principle behind electromagnets, a major component in a telephone.

Figure 9.2: Hans Christian Oersted.

3. Michael Faraday

Michael Faraday is the acknowledged father of modern electricity. His discovery is the reverse of Oersted's. While Oersted's experiment demonstrated that current-carrying conductors produce a magnetic field, Faraday discovered that a changing magnetic field produces or induces current. He demonstrated that a changing current in a coil induces current in a nearby coil even if the coils are not physically connected. This is called the principle of electromagnetic induction.

With this principle, Faraday built the world's first electric generator. The telephone system was then assured of a stable and continuous source of electrical energy. In 1831 people were producing dynamos or generators but somebody had to show them how to use electricity in communicating.

Figure 9.3: Michael Faraday.

4. Joseph Henry

Joseph Henry worked on the same experiments on electromagnetic induction as Michael faraday but he did not publish them first. He did not only show electromagnetic induction but he also showed that electromagnetism can be used to communicate with.

In a class demonstration that he performed, he showed how an electromagnet with current attracted an iron bar which in turn struck a nearby bell. This was considered a case of electrical signaling. Henry did not pursue electrical signaling, instead he helped Samuel Morse in making a working telegraph.

Figure 9.4: Joseph Henry.

5. James Clark Maxwell, Heinrich Hertz and Guglielmo Marconi

which he called radio waves to send messages without wires- the beginning of wireless communication. Maxwell put together the principles of Oersted, Faraday and others and predicted the existence of electromagnetic waves. Twenty years later,

Figure 9.5: James Clerk Maxwell, Heinrich Hertz and Guglielmo Marconi.

Hertz Experimentally showed the Existence of such Waves.

6. Thomas Edison

While Henry's use of the electromagnet paved the way for the telephone receiver, Thomas Edison devised the carbon transmitter works on the principle that the electrical resistance of carbon changes with acoustic pressure exerted on it.

Figure 9.6: Thomas Edison and Lee De Forest.

7. Lee de Forest

Long distance service was very limited. Voice signals could not be transmitted over long distances due to energy losses in a telephone line. A device was needed to energize or amplify the signals. Lee de Forest built a vacuum tube, the triode which made possible the building of a nationwide telephone system. On January 25, 1915, the first transcontinental telephone line using three vacuum tube repeaters opened between New York and San Francisco- the world's longest telephone line. This invention ushered in the electronic age for telephone.

8. William Shockley, John Bardeen, and Walter Brattain

This trio developed the solid state transistor. The transistor took over the function of the bulkier, not-so-efficient and more expensive vacuum triode in the telephone network. Later it would revolutionize every aspect of the telephone industry and all of communications.

Figure 9.7: William Shockley, John Bardeen, and Walter Brattain.

9. Donald Keck, Robert Maurer, and Peter Schultz

This trio produced the first optical losses low enough for wide use in telecommunications. The discovery was quickly recognized as a breakthrough, paving the way for the commercialization of optical fiber and literally creating a revolution in telecommunications.

Figure 9. 8: Donald Keck, Robert Maurer, and Peter Schultz.

THE LAUNCH OF ELECTRONIC TRANSPORT NETWORK

The electromagnet laid the fundamentals for large-scale electronic communications. In 1825, William Sturgeon (1783-1850) exhibited its power by lifting nine pounds with a seven-ounce piece of iron wrapped with wires through which the current of a single cell battery was sent.

Joseph Henry, an American (1797-1878), established the potential of Sturgeon's device for long distance communication by sending an electronic current over one mile or wire to stimulate an electromagnet which caused a bell to strike. This experiment was conducted in 1830. Thus the electric telegraph was born. Samuel F.B. Morse (1791- 1872), fruitfully sold Henry's invention commercially.

The telegraph was the first instrument to send messages by electricity. It was in 1835, while a professor of arts and design at New York University, that Samuel Morse attested that signals could be transported by wire. He utilized a telegraph key to encrypt messages using special code- the Morse code.

The Morse code was composed of dots and dashes demonstrating the letters of the alphabet and made by making and breaking the transmitter circuit. An electromagnet deflected the pulses of current, which moved a marker to generate written codes on a piece of paper. The instrument was improved to engrave the paper with dots and dashes. Then the operation progressed into sending by key and receiving by ear. The electric signals are sent through the wires to the receiver end where the received signals are transformed into sound by sounder. A well-trained Morse operator could transmit 40 to 50 words per minute.

Year 1877, all rapid long distance communications relied on the telegraph. That year, an enemy technology industrialized that would obtain change in the face of communication- the telephone.

THE TELEPHONE SYSTEM

The success of Alexander Graham Bell with the telephone came as direct result of his attempts to develop the telegraph. Bell was born on March 3, 1847 in Edinburgh, Scotland. He was the son and grandson specialists in pronunciation and the correction of speech. Accomplished with proper education and will pursue in the same career like his father and grandfather; his knowledge of the nature of sound led him to teach the deaf, but also to devised the telephone.

It was Bell's broad knowledge about the nature of sound and his comprehension of music empowered him to calculate the chance of transmitting several messages over similar wire at the same time. The multiple telegraph existed that time but Bell's idea was accompanied by his own musical or harmonic approach as a practical solution. The harmonic telegraph that Bell offered was based on the principle that several notes could be directed instantaneously through the same wire if the notes or signals vary in pitch. It was proven that different tones would vary the strength of an electric current in a wire. In order to have a positive result they needed to construct transmitter with a sheath which has the capability to withstand varying electronic currents and receiver that would generate these variations in audible frequencies. Bell's telephone was a duplicate of instrument through which sounds were first transmitted in 1875.

March 10, 1876 was the date that Bell achieved his greatest success because of the birth of his telephone and the death of the multiple telegraph. His work with being able to talk with electricity outweighed anything especially increasing the capability of dot-and-dash system that could offer. His invention contained communications potential that made it to flourish. The telephone is an energy converter. It transforms sound to electricity then back to sound. This transformation takes place in the mouthpiece and in the earpiece correspondingly.

The transmitter which was the mouthpiece, is actually a microphone. It contained a flexible iron diaphragm, pressing against a metal box containing carbon granules. Carbon is not a good conductor of electricity. It acts as an electrical resistor because of its loose arrangement. The resistance it offers to the current depends on how firmly the granules of the diaphragm vibrates and compresses and releases the carbon granules and change the resistance accordingly. The microphone is connected to a battery that makes the circuit current changes with the resistance and an electrical waveform is produced which closely follows the original sound waveform. The electric current travels through the wires and switches and reaches the receiver of the other telephone. The receiver is doing the opposite function of the mouthpiece, it is responsible in converting electric current into sounds. The earpiece, the receiver, is a loudspeaker. It is composed of flexible iron diaphragm near the poles of a magnet shaped in horse shoe. The varying current from the microphone passes through the coil and changes the magnetic field strength of the electromagnet. The electromagnet vibrates the diaphragm in exactly similar pattern as the microphone diaphragm vibrates. The vibrating diaphragm vibrates the air around it and recreates the speaker's voice as a sound vibration in the air.

The ability to be connected to any one of a large number of subscriber is the major feature of the telephone system. The signals leave the house to the telephone box outside. The distribution cable serving a street joins other distribution cables serving the subdivision. The local exchange is connected to all phones in the subdivision. Switches consists of the control unit, common equipment and network. The control unit composed of a computer and associated equipment. It does all the thinking in processing a call through the network. The common equipment part of the central office switch supplies services that are shared by all telephone lines. The network is a matrix of possible paths between all telephone lines. It is in the network where the connection is made. Trunk is the name for communication channels connecting two central offices. The telephone network is composed of switching centers and interconnecting trunks. The multiplexer is an instrument that combines many channels before sending out on a communication system. The system connecting switching centers may be wire, radio, and microwave, satellite or fiber optic cable. A call from one area is routed via exchanges; the number dialed informs each control unit of the direction required. Most modern exchanges use solid state electronic devices to carry out the switching process, and computers to control the operation. To do this efficiently, the voice signal must be converted to digital form or a series of pulses.

Understanding the Telephone better

A telephone is a device which is categorized as a telecommunication instrument. It permits two or more subscribers to have a conversation when they are located far from each other. A telephone transforms sound most effectively, the human voice into electronic signals qualified for transmission through wires or other transmission media for long distances and relays such signals concurrently in the form that can be heard by the users.

In 1873, Alexander Graham Bell, a Scottish emigrant, was the foremost person that acquired a United States patent for device that generated clearly, understandable replication of the human voice. This instrument was subjected to development by others. The telephone was the first instrument that allows people to talk directly to each other even they are far from each other with a large distance. The indispensable effect of the telephone had become rapidly adapted by businesses, government and households and nowadays the widespread of small appliances.

The most important parts of the telephone are microphone to speak into and an earphone which replays the voice in a far location. Most telephones consist of a ringer that generates sound to signal that someone is calling and a dial or keyboard that is used to encode the telephone number when doing a call for a person. In 1978, most telephones used rotary dial which was replaced by the modern DTMF push-button dial. This was introduced to public by AT & Tin 1963. The handset of the telephone is composed of the receiver and transmitter which held up to the ear and mouth when a conversation is happening. The dial may be located either on the handset or on the base unit to which handset is connected. The receiving telephone transform signals into audible sound in the receiver. Telephones permit duplex communication which usually means people talking on both ends concurrently.

The first telephone were directly linked to each other from the office or residence of one user. Being impractical it was replaced by a land line telephone which service is linked by a pair of steady wires to a local central office switching system. It was developed into automated system in the early 1900. Various radio system were developed to enhance mobility of the transmission between stations and automobiles in 20th century. Telephones that are hand held mobile phone was proposed for personal service in 1973. Several mobile phone networks operated around the world in 1983. Advanced Mobile Phone System (AMPS) was introduced and suggesting a standardized technology providing the users even they are far from offices or residence. These cellular system using analog progressed into digital networks with enhanced security, capacity is now greater compared to the old ones, better coverage and lower cost. The public switched telephone with its hierarchical system of many switching line has the ability to distinguish telephone number that may be called from any authorized telephone on the network.

Telephone is originally designed for simple voice communication, however as time passed, confluence has enabled most modern cellphone to have many additional capabilities. These capabilities include record spoken messages, send and receive text messages, take and display photographs or video, play music or games, surf the internet or do road navigation or immerse the user into virtual reality. Since 1999, the trend for mobile phone is smartphone that integrate all mobile communication and computing needs.

Plain old telephone service or a traditional land line telephone system usually carries both central and audio signals on the similar twisted pair of insulated wires, the telephone line. The ringer, hook switch and a dial are the three components of the control and signaling equipment. The ringer, beeper, light or other devices notifies the user to incoming calls. The hook switch signals to the central office when initiating a call. A dial is used by the user to transmit a cellphone number to the central office when initiating a call. Before dials used the almost exclusively the rotary technology which was replaced by dual tone multi frequency signaling or (DTMF) with push button telephone. A major expense of wire-line telephone service is the outside wire plant. Telephone transport both incoming and outgoing speech signals on a single pair of wires. A twisted pair lines rejects electromagnetic interference and cross talk better than a single wire or an untwisted pair.

ADVANTAGES AND DISADVANTAGES OF COMMUNICATION TECHNOLOGY TO SOCIETY

The driving force behind economic growth, citizen engagement and job creation is the technological progress. Specifically, Information and Communication Technologies (ICTs) are remoulding many features of the economies, government and societies of the world. Nowadays, it can be gleaned on different studies and context that access to the Internet has developed to be an essential tool for progress. Universal and reliable Internet access, which means that the availability of the Internet must be accessible in all areas geographically near or far, is required in a digital revolution also known as the so-called Fourth Revolution. Internet is great help in developing countries, without Internet countries won't be able to wholeheartedly contribute in a progressively mobile and digital-based economy. Accessible Internet has become one of the pillars of country who are aiming for development and for making its nation to be globally competitive. Communication among different countries through Internet strengthens the relationship and avoid conflicts.

Government officials, businesses either big or small and citizens in progressing countries can develop the metamorphose power of ICTs to make available more efficient services, modify economic growth and strengthen social networks. The ninety-five percent of global population now live in places that are covered by a mobile-cellular networks. In Tanzania, for instance, the agents of mobile money are now outnumbered all other financial intercessors by a factor of 10 to 1. Meaning more than half of the population of Tanzania who are living on less than $2 a day can access the mobile technology. However, though the digital revolution is a global phenomenon, there are still large inconsistencies between and within the countries when it comes to the speed, penetration, affordability and performance of ICT services. But even though the digital revolution is a global phenomenon, there are still huge disparities between and within countries when it comes to the penetration, affordability, and performance of ICT services. Despite the fact that almost half of the population of the world in the year 2016 had said to be that they can access the Internet but the penetration rate or those who can really use Internet services in the least developed countries was only 15% whereas in seven individuals only one person can really penetrate or surf the Internet with full access because of the issue on its signal of being available. One of the factors that contribute in accessing the Internet is that the internet through mobile or fixed broadband still dwell in excessively expensive in many progressing countries, where ICT infrastructure is lacking and regulatory blockage hinder the development of broadband. The charge of mobile-broadband services as of December 2015

One contributing factor is that access to the Internet through mobile or fixed broadband remains prohibitively expensive in many developing countries, where lack of ICT infrastructure and regulatory bottlenecks hamper broadband development. As of December 2015, the cost of mobile-broadband services amounted to approximately 17 % of the average monthly Gross National Income (GNI) per amounted to about 17% of the average monthly Gross National Income (GNI) per principal investment in the developed countries, there was a huge gap of discrepancy if it was compared to just 5% globally. The swiftness or speed of broadband services also varies, they always differ from time to time or base on the country where you live in. Fixed-broadband speeds of 10 Mbit/sand higher are ordinary in developed countries; against with the LDCs which only gets 7% of fixed-broadband services reach 10 Mbit/s. Under SDG #9, it is conceived that the world has set a determined aim that is to significantly upsurge the access to information and communications technology and struggle in order to give universal and reasonable access to the Internet in somewhat developed countries by the year 2020. It is perceived clearly that the government, the private sector and the international community have a lot to prepare in order to reach the target and bond the digital gap.

However, development is achievable. Efficient policy improvement on ICT can generate private investment in broadband infrastructure and provide Internet access more affordable. Fair taxation can be mandated by the government for the telecommunication industry and use worldwide service funds to concentrate on rollout of the broadband, in cooperation with others and under open entrance principles to improve healthy competition. Poverty reduction and shared prosperity was supported by these efforts directly.

Information- Communication Technology (ICT) is a common manifestation for a variety of different computer, information and communication devices, applications, networks and services. Communication Technology has been a vital part of our daily lives. Everybody both individuals and the sectors of industry use communication technology in order to acquire what they need to make their lives easier or for a certain function that technology may offer. But bear in mind that these applications have their downsides as well. Voice mail, Email, Teleconferences and Compressed video, GDSS (Computer assisted decision making) and Virtual reality are the common technology that has been involved in an organization.

Figure.9.11: The mobile phone.

All these types of communication can contribute in the growth of an organization, however they also have their limitation and downsides.

Here are some advantages and disadvantages of Communication Technology.

ADVANTAGES OF COMMUNICATION TECHNOLOGY IN AN ORGANIZATION

1. Swiftness in sending and receiving information

Electronic mail and text messaging systems are communication technology that aim to speed up the sending of information within and outside the organization. Sharing of information within the organization with the use of decentralized computing systems is very fast because all the data are accessed from one central unit called "database" and it is collectively distributed across different departments within the organization.

2. Advances in organizational communication

The design of a shared information environment in an organization was provide by communication technology. All the information of the organization are organized in one central location, permitting every member to access information that they need. Electronic mail technology will open communication between low level employees with the manager without any need of having a committed time for meeting. This permits information to current easily from bottom to the top without any hindrance.

3. Faster decision making in an organization

Due to the faster transfer of information through communication technology, employees in an organization can easily ask each other and analyse information in a shortest period of time and create s decision. With the utilization of OLAP or online analytical processing employees can conduct inquiries on a database to access data about a particular client or matter and decide quickly on decision based on the data that can be retrieved from the database about a particular client or matter.

4. Participation in organizational processes increases

All the members of the organization had assigned special tasks to complete and this can only be observed at all by the manager. All will be active since their work has been made easier by easy communication. Messaging services to request help from their co-workers or workmates within the organization, communication technology is of a great help in using different Medias that technology can offer in an organization and participate in decision making and creative design challenges.

5. Affect the way people interact in organizations

Let us cite an example of situation where communication technology is at great help in your work. The operations manager visits the site twice a week, by the period he comes, most of the employees want to talk with him and he will have limited time for each of you. However by utilizing electronic email, you can just draft an email and send it to your manager or workmate. In this situation you don't have to stop your work or what you are doing to ask queries and they don't have to stop what they are doing to attend to your queries because you will get instant replied and continue your work while having a conversation with the manager.

6. Arrange organizational life

Communication technology clean all the things that needed to be cleaned, they put everything where these things supposed to be. There is no need to see papers and files everywhere in the office. Data are stored on the database and employees will have less paper work in the organization which makes their work a lot easier. Most importantly they can spend their time on more productive activities or section since the time to organize their data is saved by the technology.

7. Establish open discussions

Organizations nowadays have utilized the internet to make organizational forums where employees or members of the organization can converse about different issues and acquire a solution. In this situation members have the chance to suggest on areas where development is needed within the organization. The discussion on forums needs no meeting at all because it can be done at your convenient time and place.

8. Delete stereotypical classification

Stereotypical, according to Urban Dictionary, is a term utilize to describe all people of a particular belief into a denoted negative category that may only perceived on few selected of the racial demographics. Through the help of communication technology, all of this will be deleted because there is no face-to face communication. Stereotyping of employees can be of negative result such as poor performance and lack of self-esteem amongst the employee, therefore it is not a good behavior.

9. Serve as a voice to those who normally would not speak up in the groups

Communication that you have to talk face to face with other is sometime has the tendency to be difficult to other people. Therefore the use of communication technology in the organization may help them to communicate effectively at work because they don't have to face other people. Some employees decide not to talk because of this fear but these employees might have relevant ideas or suggestions that can make or contribute to the progress of the organization so they can easily use technological tools like electronic mail to speak out their mind.

DISADVANTAGES OF COMMUNICATION TECHNOLOGY IN AN ORGANIZATION

1. Employees don't get a chance to build rapport to each other.

Communication technology such as electronic email is a poor substitute for face-to-face communication. Employees are utilizing machines to communicate to each other therefore they have no time talk to each and get to know each other. The outcome is an increased with bad relationship at work because they do not have the chance to build rapport to each other. Let us say that these shy employees will be benefited to this kind of working environment but the chance to grow while mingling with others is stolen from them. They won't learn how to interact with others and may keep them to hide all the time from many people. Actually this situation is not a great help to those shy employees, they are just being drag far more deeply in loneliness.

People are very far from machines. In every organization, members are given a chance to meet and interact with his co-workers in order for to build rapport or harmonious relationship toward each other. Love and concern must be felt to each and everyone in the organization because when employees work in a happy working ambiance, his performance will be automatically good. Actually every organization allots a certain amount of money for the team building or rest and recreation activities of the employees for them to build love and concern to each other. Utilizing the tool of communication can also be done if some announcement can be posted on it like group chat., where all the employees are member of it and they can access to all the information or announcement that are being posted on it. However, this is also not suitable to some organization especially the access to internet is very limited. Announcements of essential information through communication technology tools can also have loopholes such as the employee is not accustomed to that website or the employee doesn't have an email account or does not have an access to expensive internet connection.

2. Expensive Training for Employees

If a new policy will be adhered in the organization, there is a need to train each and every employee. Aside from the expenses of acquiring such communication technology tool, another expenses for the training of the employees. There are also instances that employees, especially the old ones who have the difficulty in learning modern gadgets and the likes, failed to learn how to utilize the machine another expenses for follow trainings or worse it can lead to job loses.

3. Expensive

Installation of new communication technology system in a very big organization is can be very expensive. For example if an organization has 50 employees and they are instructed to get the data from a centralized location. They will have to build a strong database which will cater all the queries made by these employees in a minute as well they have to buy a computer for each of the employee. Acquiring all these computers demand for a centralized air-conditioned room in order for the computers to have long functional life span.

4. Centralized database is not secured.

If the employees will get information for the database all will be authorized to manipulate the database which will make it exposed to people with wrong intention within the organization. It can be hacked by hackers and get all the essential information of the organization. If the database will be attacked by a hacker or virus all information might lose in just a minute. Therefore the organization have to allot another expenditure for the maintenance of the security of these information.

THE IMPACT COMMUNICATION TECHNOLOGY IN PEOPLE'S LIFE

Remain in Contact

A century has passed, progress in technological aspect have conveyed the world closer, creating communication through long distances possible and a lot easier. Telegrams reach its destination faster than letters; however phone calls are transported faster than telegrams. Phone calls are easier and more enjoyable because they require no hindrance and allow each other to hear each other's voice. Another technology that made communication a step further is the cell phones. It allows people to call and talk to each other heedlessly of their location. Communication using internet is the most efficient but with electronic mail being the most related type of the paper letter version. Webcams which is paired with communication systems such as Skype, iChat or Google Video Chat, make it possible to see the person who are talking to regardless of the distance that separates you. You could not only hear the voice but you can also see the person you are talking to.

An Aid to Business

Similar technological development that have abridged and enhanced personal communication are also bring benefits to the field of business. Keeping in touch through technological advancement between colleagues is possible whenever they are just near or a country apart to each other, companies use video conference to have their workers hold efficient meetings and discussions even though they are scattered around the globe. Social media is responsible for business networking and online networks planned specifically for that purpose, such as LinkedIn. Companies can enlarge their local market and obtain a wider customer base simply by maintaining an active online presence.

Overpowering Disabilities

Disabled people are one of the most beneficial of the communication technology. It has improved communication for disabled people and made it possible even though they were born with such defects. Hearing aids increase hearing of partially deaf people, allowing them to understand speech easier, while the completely deaf can hear through cochlear implants. Cochlear implants can restore hearing to the completely deaf person. People with severe speech impairments have their way to express themselves through speech-generating devices that were designed just for them. Stephen Hawking who is a scientist is a famous user of such device. Progress in technology may result in functional brain-computer interface systems, renovating the ability to interconnect to people who have lost it completely, such as those who suffers of locked-in syndrome.

Attainment a Broader Audience

The range of messages widens as the people's ability to converse progressed. Politics and activism require its members to be exceptionally good in conversation because that is their capital in attaining listeners and ally. For example, photos and video which are recorded secretly through a cell phone can be speedily and easily shared online through websites such as You Tube, creating it difficult to repressive government to keep control; social networks such as Twitter and Facebook can be utilize to organize and manage meetings and protests. The revolution of Egyptian in year 2011-2012 was encouraged greatly by social media. Communication technology a technology that permits communication has aided to transfigure the communication process. Automated electric communication, like text messages, email and social networking websites, has allowed society to become more associated.

Positive Encouragements

Organizations utilize electronic communication in day-to-day processes, and everyone uses it for personal tasks, like banking, paying bills and socializing. Communication technology has also created long-distance communication and mass communication easy and convenient.

Social Influence

Technology that majors in communication, whereas Internet communication such as networking websites and dating websites, has made the operations of socializing has been made simply communicable, Social networking websites permit you to easily stay in touch with friends, however dating websites will provide you someone who has the highest percentage of being handsome will help you to find someone compatible to being a relationship with.

Communication technology, especially Internet communication like social networking websites and dating websites, has simplified the process of socializing. Social networking websites enable you to easily keep in touch with friends and family, whereas dating websites help you find someone compatible to be in a relationship with.

Mass Communication

Communication to a large number of people is called mass communication. Schools and business are two types or organization which utilize electronic communication, such as email and e-newsletters in order to communicate to a large number of people.

Long-Distance Communication

In ancient times people utilize pigeons in delivering their messages. Incredibly served as the runner of two persons who are prohibit or not allowed to see each other by simply sending a letter using the pigeons. This long distance relationship ran messages through sending pigeons back and forth, they trained these pigeons in order to fly written messages. The progress of communication technology has. The advancement of communication technology has assisted to create long-distance communication fast and long-distance communication fast and apparently uncomplicated.

Communication Action for the Disabled

The beneficial that can give by communication technology is not often talked about the technology's ability to give disabled people the ability to communicate. Many had pledge their help and willing to extend in buying essential procedure in order to strengthen the voice or noise that the person who have gone unheard because of his illness.

CHAPTER 10

WIRELESS COMMUNICATION TECHNOLOGY

INTRODUCTION

Society has been very lucky for all the things that have been invented by the scientists that can make our life progressive. They endure all the hard works just to solve natural phenomenon that can make life a lot easier compared to the ancient times. Modern people owe it to the scientists of ancient times. Life has become better and more progressive. It will be the responsibility of the new generation to continue the track that great scientist had created. To the students of today who will use all the knowledge that those great people had contributed, the future of tomorrow is in your hands. Appreciate and utilize the legacy that they have left to this world. Especially to those parents who give and endure all the sacrifices that we can do to our children in order for them to realize that having access to Internet can be both beneficial and perilous.

The effort made by the scientist in order to create things that can be useful to mankind. Just like the wireless communication where in you can bring your cell phone anywhere and take calls and text message while strolling in a far places.

Today our society has become increasingly reliant to communication networks. Therefore, engineers and scientists continue to search for more dependable and efficient communication systems. At present scientist wants to exploring the possibility of using laser light that have been used rather than radio waves as a carrier of information across the world and beyond.

LAUNCHING OF WIRELESS COMMUNICATION

Heinrich Hertz, a German physicist (1857-1894), conducted an experiment in 1886 which concreted the way for another innovation in terms of communication. His work is consisted of a set-up which is composed of two circuits. Circuit 1 had a small gap and he connected it to a high-voltage source. Whenever a great amount of voltage had built up in the gap, sparks jump back and forth across the gap. Circuit 2 in the experiment of Hertz is consisted of a simple conductor bent into a circle with a small gap between its ends. Upon the experiment, Hertz perceived that every time that the circuit 1 sparks, sparks will also be perceived in circuit 2 even though there is no wire that connects them. He assumed that energy from the spark of circuit 1 was generated through the air and acknowledged by the second circuit. The energy was transferred from circuit 1 to circuit 2 through electromagnetic waves.

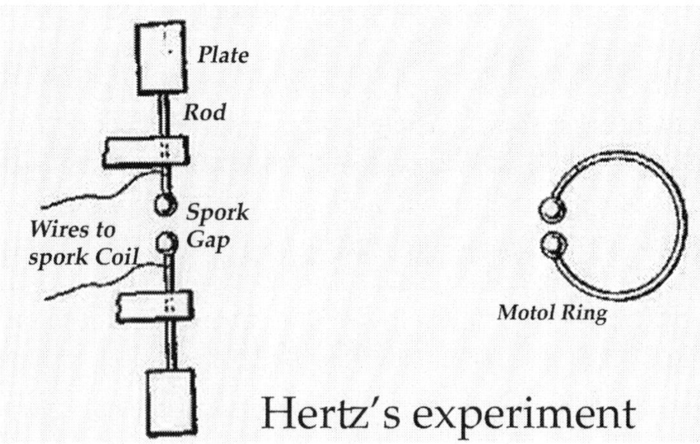

Figure 10.1: Experiment set-up conducted by Hertz.

Unfortunately Hertz did not realize that this ability to transfer information without connecting wires would be an applied method of communication. However, Guglielmo Marconi perceived this ability and devised the first wireless telegraph. Evidently, using a spark gap transmitter and a simple receiver for the indication of signal by clicks in a telephone earpiece, he had sent messages a few kilometers. In 1901 he used the same method to collect Morse signals across the Atlantic.

Electromagnetic waves as message carriers utilize by the modern telephone system for transoceanic communication. Radio and satellite communication are also wireless. They function without connecting wires between the transmitter and the receiver. It is evident that communication happens through electromagnetic waves.

ELECTROMAGNETIC WAVES

James Clerk Maxwell, the Scottish physicist (1831-1879), created equations merging the theories of electricity and magnetism. The equations that he have created characterize a wave composing of changing electric and magnetic fields. Moreover, his formulas suggested that such a wave transmit at a speed of 3×10^8 m/s in a vacuum just like what light's behavior. An electric body is enclosed by an electric field. When the charge is not moving or at rest, the field at a point be contingent on its distance from the charge and decreases in strength at points far from the charge. When the charge is in motion, the field it carries changes with time and space.

Wireless communication is the transmission of information between two or more references that are interconnected by an electric probe. Wireless communication is commonly associated with telecommunications industry which is refer to telecommunications systems such as transmitters of radio, receivers, remote controls and many others that utilize some formula of energy to transmit information even without the use of connecting wires. Information can be transmitted in both near and far distances. Operations regarding with wireless communication permit services like long-range transport network that are otherwise unreasonable to device with the utilization of wires.

The electromagnetic wireless telecommunications is most commonly adapted by most common wireless technology such as radio or infra-red signals. In infra-red waves can transfer signal within a short distance only such as the remote control of the television while radio waves can travel as remote as thousands or even millions of kilometres for deep-space radio communications. It includes different types of constant, transportable and portable applications which includes radios, cellular telephones, personal digital assistants and wireless networking. The applications of radio wireless technology includes other examples such as GPS units, garage door openers, wireless computer mice, keyboards, headphones, satellite television, broadcast television and cordless telephones. Less shared approaches of attaining wireless transport network include the use of light, sound, magnetic or electric fields.

Mobile or cellular phone is one of the famous and essential examples of wireless technology. There are more than 4.6 billion mobile cellular subscribers on earth entirely as of the end of 2010. Cell phones a wireless device utilize radio waves to allow their subscribers to make phone calls from many locations worldwide. They can utilized contained by a range of the mobile telephone sites that house the essential tools to transport and receive the radio signals that these devices release. Wireless data transport networks are also an important part of mobile computing. These different existed technologies varied in local obtainability coverage range and presentation. In some occurrences, subscribers must be able to implement several connection types and adjust between them. To make the experience of the subscribers simply, connection manager software is available, or a mobile VPN can be used to manage the multiple connections as a protected, on a particular virtual network. Wi-Fi is one of the popular supporting technologies, this is a wireless local area network that empowers portable computing devices to connect easily to the Internet. Wi-Fi, standardized as IEEE 802.11 a,b,g,n, deals with the speed of some kinds of wired Ethernet. Wi-Fi has become the most significant way to connect to Internet and given the term the de facto standard for access in private home, within the offices and even in public places that offers hotspot. Many of the businesses charge people with their access to the Internet monthly and others offer it as free in order for them to encourage customers that will patronize their goods or services.

Cellular data amenity offers coverage enclosed with a range of 10-15 miles from the nearest cell site. As technologies have advanced speeds have increased as well, from earlier technologies such as GSM, CDMA and GPRS to 3G networks such as W-CDMA, EDGE or CDMA2000. Mobile Satellite Communications may be used where new wireless connections are inaccessible, like in huge rural location or remote areas. Satellite communications are extra essential for transportation, aviation, maritime and military use.

WIRELESS COMMUNICATIONS

Radio Waves

Radio waves are utilized for communications over radio and TV programmes. Everyone who has an earpiece can tune it to the radio frequency to receive the signal. If the radio stations use the same broadcasting frequencies the waves sometimes obstruct with each other. Ionosphere which reflected medium wavelength radio waves is used for long expanse communication, but not for interconnecting with satellite directly above the ionosphere.

Microwaves

The transmitter and receiver must be visible to each other that is why they are located in high, usually on tall posts. They have to be located together so that hills, or other high landmasses cannot hinder the beam. Signals are transferring to and from satellites, which then convey signals all over the Earth. This occurrence takes place when transmitting information for TV programmes, conversation over the phone, or observation of the Earth especially in forecasting of weather that is taking place with variations depending on the data perceived.

Light, a Part of the Big Family

Current studies showed that light and radio waves belong to a big family of electromagnetic waves. The family of waves is denoted to as the electromagnetic spectrum. The spectrum is divided into several groups of waves. Each group is described by a range of occurrences or wavelengths. The waves are categorized conferring to their process of propagation.

However, regardless of the frequency of the waves, they all have common properties.

1. They all transmit at the same speed in free space (3x 10^8 m/s).

2. They display diffraction and interference phenomena.

3. They follow the laws of reflection and refraction.

4. They can be polarized.

Figure 10.2: Satellite communications which relay signals back and forth the Earth.

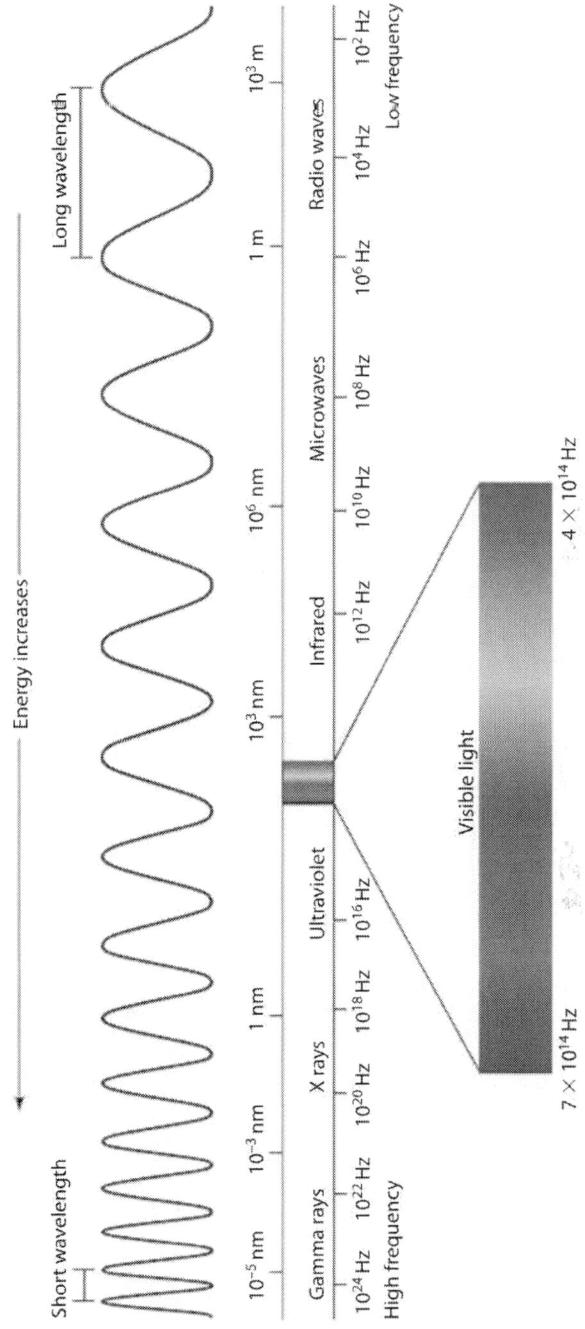

Figure 10.3: The Electromagnetic spectrum.

Various areas of the electromagnetic spectrum have varying characteristics.

Table 10.1: Characteristics of the different regions of the electromagnetic spectrum

Name/ Group	**Method of generation**	**Remarks**
Low-frequency radio wave	Oscillation in electric circuit	Radio waves are seldom used below 10, 000 Hz. This group includes weather and some aviation broadcasting.
Broadcast radio wave	Oscillation in electric circuit	This band of radio waves has been set aside for the regular AM broadcast band.
High-frequency radio waves	Oscillation in electric circuit	This region has many kinds of broadcasting amateur radio, TV, FM radio and experimental work.
Infrared	Vibration of atoms within molecules; molecular vibration in solids	This is known as radiant heat emitted by very hot bodies.
Visible	Motion of electrons in atoms	This is the band to which our eyes are sensitive.

Ultra-violet	Motion of electrons in atoms	Exposure to it stimulates the production of Vitamin D in the skin. It is also responsible for sunburn.
X-ray	Rapid acceleration or deceleration of charged particles	It is used in medicine and detects flaws in metals and plastics.
Gamma ray	Nuclear reactions, radioactive disintegration	It is very energetic and highly penetrating.

Diffraction

Diffraction refers to the dispersal of a beam through holes and around corners. The extreme effect perceives when the hole has the same size to the wavelength. Radio waves of approximately 5 m are deflected by big buildings. Radio waves of1 km are deflected around mountains and through valleys, so they are able to arrive to most locations and are appropriate for communications.

The transmitter and receivers must be in route in line with each other because microwave beams of a few centimetres do not disperse around junctions or around the hills. In order for the microwaves to be transported from satellite dish the wavelength must be shorter compared to the diameter of the dish to decrease deflection. Therefore, if it to be compared with radio waves, microwaves can be conveyed as a thin beam.

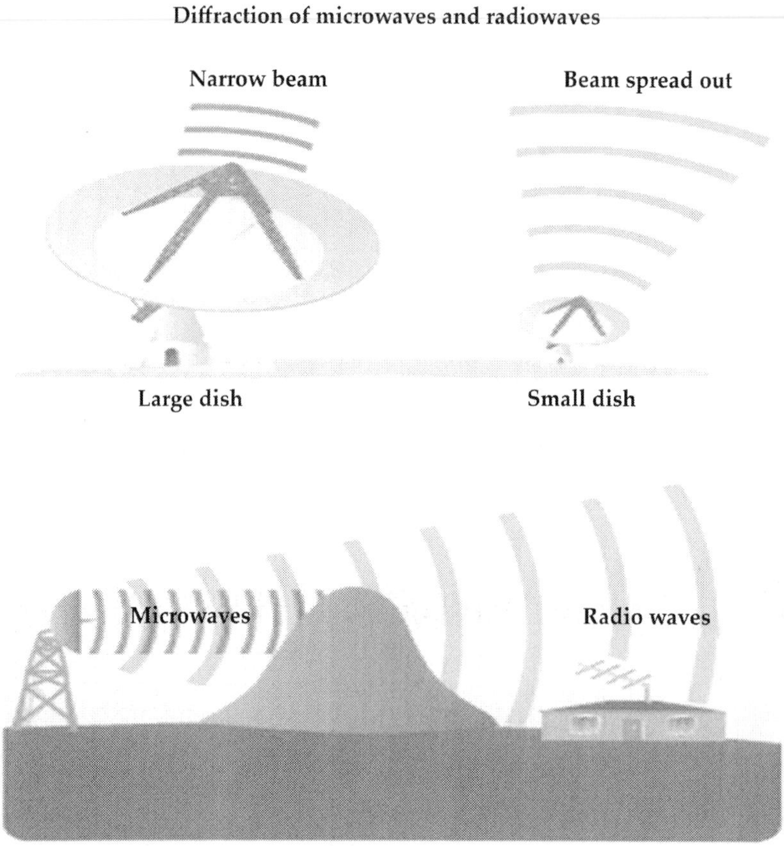

Figure 10.4: Diffraction of radio waves and microwaves.

Analogue and Digital Signals

Electromagnetic waves such as radio waves, microwaves, infrared and visible light are being utilized to convey information, whether these information are composed of sound, pictures and other data. The information conveyed by these electromagnetic waves are referred to as the signal. Carrier wave is added to an electromagnetic wave to make it available for transmission. After transmission as the wave is received the carrier wave will be eliminated and the signal is reconstructed. Signal is categorized into two namely analogue and digital. Ah analogue signal transforms in frequency and/or amplitude continuously in a manner that the changes in the voice or sound being transported will be identical.

Meanwhile, digital signal is composed of two values, denoted as 0 and 1 (or on and off). The signal is transformed into code of 0s and 1s. It develops a stream of 0 and 1 values. These streams of pulses then added to the carrier wave and transported. When the signal is received, it is interpreted to recover the original signal.

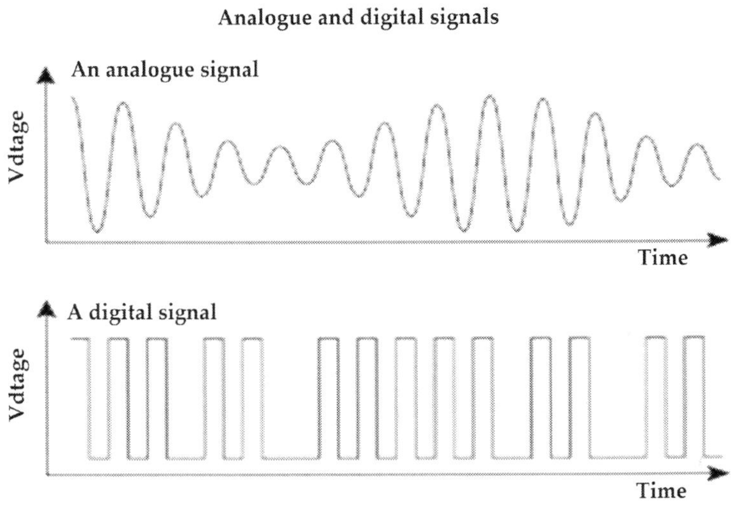

Figure 10.5: The analogue and digital signal.

Wireless and Mobile Phones

In order for the information to be transmitted wireless communication utilizes microwaves and radio waves. This kind of transmission has offered several advantages. The advantages are as follow:

1. Cable wires are not needed to connect laptops to the internet, or for cellular phones or radio. No wires are needed to connect laptops to the internet, or for mobile phones or radio.
2. Making phone calls and sending electronic mails are available limitless within 24 hours a day.
3. Convenience will be observed in communication using wireless technology like because of its portability you can bring your gadget anywhere provided signal is available and you can access the Internet and do your work.

Mobile Phones

Microwave signals are used by mobile phones. Transporting signals from mobile phones are replicated by metal surfaces and walls, then transmit through the air to interconnect with the nearest transmitter post. Network of transmitter posts are present to convey the signals on to the nearest post to the receiving phone. It is not reported if the long-term of using mobile phones has long-term effects to the users. However, there is a possibility that the residents who live near to mobile phone posts are subjected to risk. It is said that technology will be accepted by the society as long as the benefits outweigh the risk, people reflect about the risks and benefits of using a mobile phone and decide to experience the convenience that the mobile phones give.

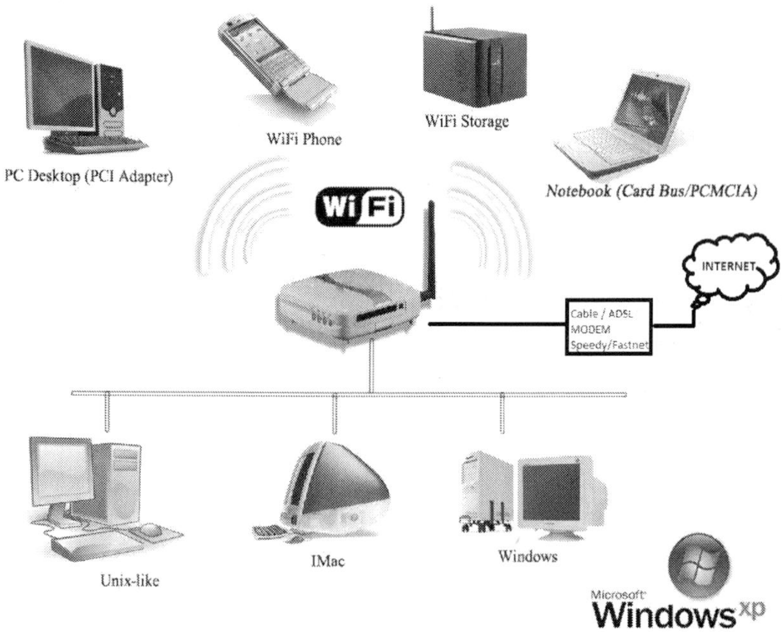

Figure 10.6: shows how wireless technology works.

Connection between Factors

A huge sample of thousands of people must be used in order for the study to be accepted as a good study. The sample will be complemented, in order for the scientists to relate similar type of people who use mobile phones with people who don't, to test if this a factor that heighten the risk. For example, they might compare women, children, or men who are exposed to other sources of radiation such as television or computers. If the result contains a correlation, they will study to see if this is due to different cause. A correlation is defined as the link between two factors. The presence of correlation does not always suggest that one factor causes the other. Like for an instance, there is a correlation between increased sales of mango and an increase in case of diarrhea, but it does not suggest that eating mango can cause diarrhea. Both increase because of hot weather. It is very essential to scientists to search an instrument that explains why a factor causes an outcome. It can be seen in a case where the mechanism by which radioactive materials cause cancer is the radiation with ionising atom and destroying the cells of the body.

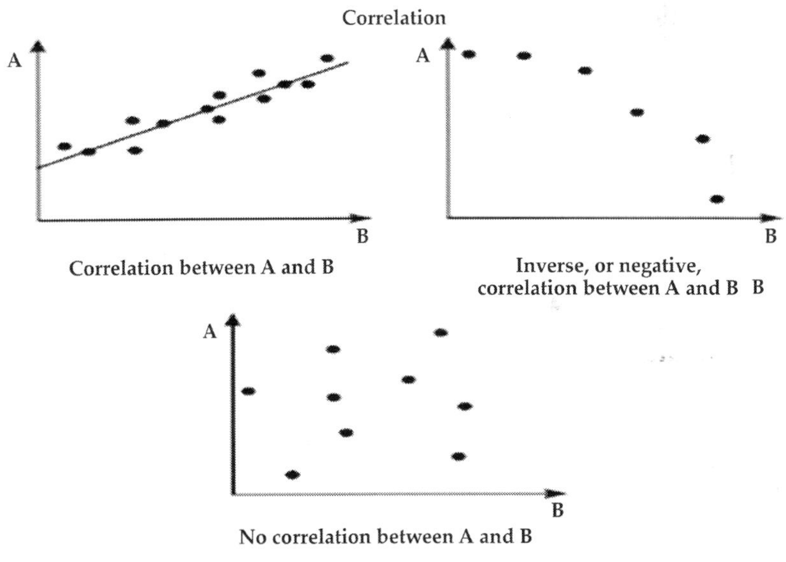

Figure 10.7: shows if there is correlation between two factors.

RADIO TRANSPORT NETWORK

Radio transport network conveys a function that includes telecommunications network, electronic system of links and switches, and the controls that govern their operation. These permit data to transfer and exchange among multiple subscribers. If several subscribers of telecommunication wish to interconnect with one another, they must be connected into some form of network. In this scheme, each subscriber can be given a straight point-to-point link to all the other subscriber in what is called as fully connected topology which is associated with the connections engaged in the earliest days of telephony, but in employing this technique is unreasonable and expensive- particularly for a large and spread network. Moreover, the technique is incompetent since most of the links will be shiftless at any given time. Modern telecommunications networks inaugurate a linked network of switches or nodes, which such that subscriber is connected to one of the nodes, to avoid those issues. Communications channel is the network that each link has. Communications channel used wire, fibre optic and radio waves for transmitting information.

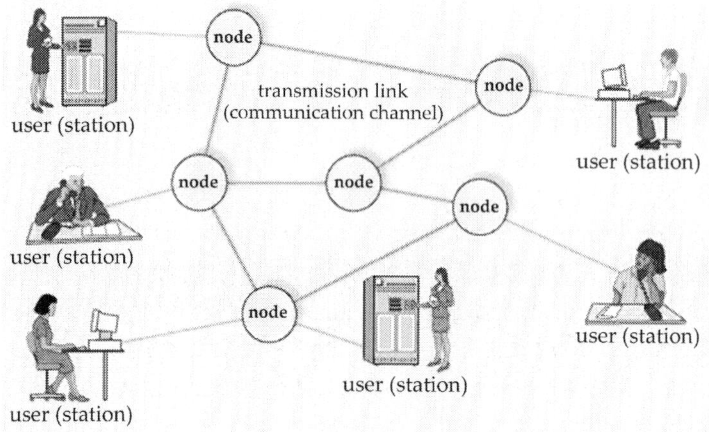

Figure 10. 8: A simple closed telecommunications network.

TYPES OF NETWORKS

1. Switched communications network

Communications network that transmits information from source to destination through a sequence of network nodes is called switched communications network. The process of switching can be achieved in two ways. A steadfast physical path is recognized through the network and is held for as long as communication is essential, this network is called circuit-switched network. The traditional telephone system is the example for this type of network. In packet-switched network is another way of switching process, routes digital information in tiny pieces called packets. Each packet continues independently through the network. The process where each packet is shortly stored at each intermediate node is referred to as store-and-forward process. Then the information will be forwarded as soon as the next link becomes available. However, in connection-oriented communication scheme, each packet takes similar route through the network, therefore arrive at the destination in the pattern in why they were directed. On the contrary, in connectionless or datagram scheme, each packet may take a dissimilar path through the network. Subsequently, datagrams may not reach its destination in the order of pattern which they were sent, they are arranged with numbers so that they can precisely reassembled. This method is utilize for transporting information through the Internet.

2. *Broadcast Network*

A broadcast network evades the extensive routing procedures of a switched network by guaranteeing that each transmissions of nodes are received by all other nodes in the network. In result, a broadcast network has only a single communications channel. A wired local area network or LAN can be set up as a broadcast network. One subscriber is connected to each node and the nodes usually arranged in a bus, ring, and star topology. Radio or optical links broadcast nodes that are connected in a wireless LAN. Most satellite radio systems, which is on a larger scale, are broadcast networks. Earth station that are programmed within the system can stereotypically receive all messages deployed by a satellite.

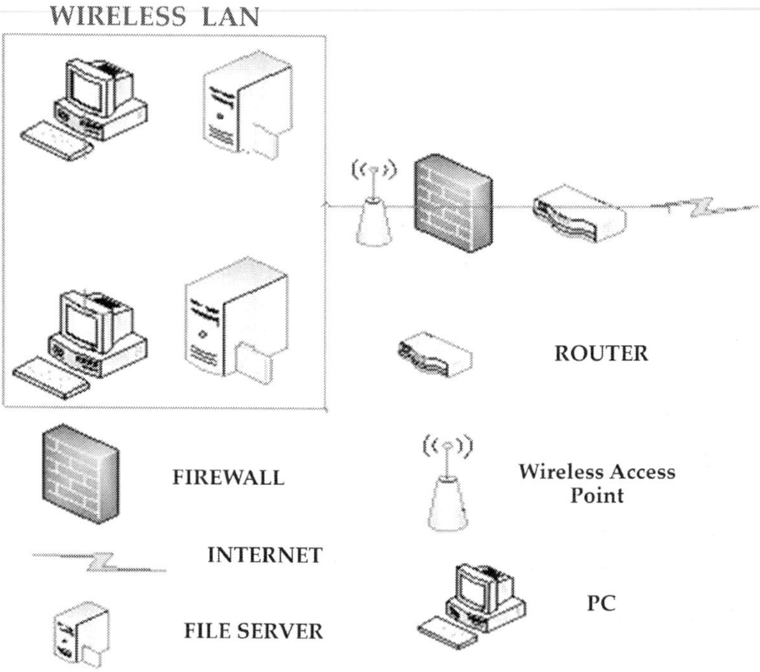

Figure 10.9: A wireless LAN set-up.

3. Network Access

As all odes can receive each transmission in a broadcast network, a process must be recognized for allocating a transport network channel to the node or nodes that consist of packets that can transport and at the same time avoid obstructive interference from crashes or simultaneous transmission. Multiple access is term used for this type of communication, wherein it can be recognized either by arranging a process in which nodes yield turns in transporting in a systematic approach or by random access to the channel.

4. Scheduled Access

Time-division multiple access (TDMA) is a scheduling method wherein a time slot is allocated in turn to each node, the slot will be utilize if it has an information to transmit. TDMA will be unproductive if some nodes are a lot busier than others because no information are accepted during time slots assigned to silent nodes. If this case occurs, a reservation system may be employed, in which there are fewer time slots than nodes and a node reserves a slot only when there is need for transmission. A difference of TDMA is the method of surveying, in which a central controller asks each node in turn if it involves channel access, and a nod transports a packet or message only in response to its survey. "Smart" controllers can reply enthusiastically to nodes that abruptly become very busy by surveying them more often for communication. A dispersed form of surveying is known as token passing. Therefore, a special "token" packet is distributed from node to node. Only the node with token is authorized to transport; then all others are listeners.

There are several disadvantages that scheduled access schemes have, these advantages include; large overhead mandated for the reservation, polling, and token passing processes and the chance of idle periods when only a few nodes are transporting. This resulted to widespread delays in routing information, particularly when heavy traffic occurs in various parts of the network at different times. It is conceived to be the properties of most practical communications networks. Random-access algorithms were created particularly to give nodes with something to transport more rapidly access to the channel. Under random access, channel is very susceptible to packet collisions therefore different processes have been advanced to minimize this chance of collisions.

5. Carrier sense multiple access

Carrier sense multiple access (CSMA) a one random-access method that minimize the possibility of collisions. In this system a node snoops to the channel first and interrupt transporting when it perceives that the channel is busy. Due to the postponement in channel generation and node processing, it is likely that a node will inaccurately perceives a busy channel to be shiftless and will cause a collision if it transmits. In CSMA, however, the transporting nodes will distinguish that a collisions has happened: the particular destinations will not recognize receipt of an effective packet. Each node then waits a random time afore sending again which anticipates to prevent another collision. This system is commonly engaged in packet networks with radio links, such as the system the system used by amateur radio operators. It is essential to reduce the time that a communications channel expends in a collision state, subsequently this effectively shuts down the channel. When a node can concurrently transport and receive typical probable on wire and fibre-optic links but unfortunately not on radio links), then it can dismiss from sending instantly upon sensing the build-up of a collision, thus moving the channel out of the collision state as soon as possible. This process is known as carrier sense multiple access with collision detection (CSMA/CD), a structure of the famous wired Ethernet. As collisions are so damaging to network presentation, various techniques have been employed to permit multiple communications on a broadcast network without inevitably causing mutual packet damage. Spread-spectrum multiple access (SSMA) is one of the most efficacious method. In SSMA concurrent transmissions will cause only a minor increase in little error possibility for each subscriber if the channel is not too seriously overloaded. Error-free packet can be gained through using an accurate control code. Weaknesses of SSM include broader signal bandwidth and grander equipment cost and complexity compared to traditional CSMA.

6. Open Systems Interconnection

Various communication requests different network solutions, and these various network practices can make a weighty problems of compatibility when networks are interconnected with one another. To overcome some of these interconnection problems and issues, the open system interconnection (OSI) was permitted in 1983 as an international standard for communications architecture by International Organization for Standardization (ISO) and the International Telegraph and Telephone Consultative Committee (CCITT). The OSI model, as shown in figure, is composed of seven layers, each of which is chosen to accomplish a distinct purpose at a different level of perception. The three layers at the bottom provide for the well-timed and accurate transmission of data, and the top four confirm that arriving data are distinguishable and advantageous. However, while all seven layers are generally necessary at each subscriber's location, only the bottom three are normally engaged at a network node, since nodes are concerned only with well-timed and accurate data transmission from a reference to another reference.

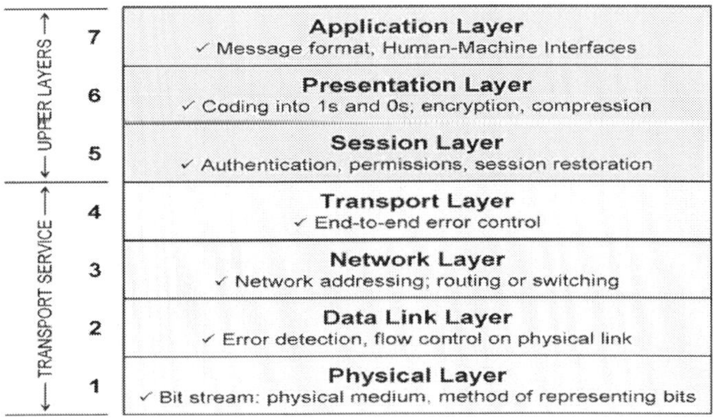

Figure 10.10: The OSI Model.

7. Data Recognition and use

Since the content of the application layer is precise to each subscriber, it is then challenging to generalize. For an instance, distributed database used in the banking and airline businesses require numerous access and security problems to be solved at this level. Transparency of network which is making the physical dissemination of resources immaterial to the human subscriber also is handled at this level. The presentation layer, on the contrary, accomplishes functions that are requested satisfactorily often that a universal resolution is reasonable. These purposes are often positioned in a software library that is available by several subscribers running different applications such as text conversion, data compression and the data encryption. Subscribers interface or boundary with the network is conducted by the session layer, which manages the process of interconnecting to another computer, confirming subscriber's legitimacy, and founding a dependable communication method. This layer also confirms that files which can be changed by several network subscribers are kept as mandated. Transport layer recognized the information from the session layer. Transport layer separates the information stream into smaller units, if needed, the transport layer may establish that all arrive correctly at the destination. If fast quantity of information is needed, the transport layer may organize several concurrently directions in the network and send various parts of the information over each path. On the contrary, if small cost is a prerequisite, then the layer may time-multiplex several subscribers' statistics over one track through the network.

Transport layer also regulates the flow control at this level, confirming that information from a fast source will not clogged a slow destination.

8. Data Transfer

The network layer disrupts information into packets and regulates how the packets are transmitted within the network, it also denote which nodes will check packets for errors along the route, and if congestion control is needed in a heavily loaded network. The data-link layer converts a raw communications channel into a line that seems principally free of conduction faults to the network layer. This is done by disrupting data up into data frames, transporting them successively, and handling acknowledgement frames sent back to the source by the destination. This layer also employs frame boundaries and administers procedures from missing, smashed, or reproduced frames. The physical layer is the transmission medium itself, along with different electric and mechanical stipulations.

COMMUNICATION AND ITS ENERGY CONSERVATION

Environment and Energy conservation is an essential topics to building owners and managers, however the test of getting occupants to buy-in to your creativities and actually change their behaviours remains difficult. Moreover, everyone is seeing the messages in the media and making modifications at home, however, they often have a difficult time linking it with things they can do at work. In order for other people alter the way they behave is to illustrate them how it impacts them personally. At home, conservation of energy is just easy to do and the result is you have to pay less. At work, these rewards don't covey much weight because the occupant is not affected by the expenses of operating the buildings they work in. This suggests that the message and the communication methods must be focused to things that matter to your spectators. Meanwhile, the occupant is not affected by the expenses, the environmental impacts give the best encouragement for change. There are effective ways to communicate the environmental impact of the wireless communication, it's just that one of the best way is to provide statistics and comparisons that the everyday person can associate to. The latest commercial by a major city about their blue box recycling program went yonder expenses and recognized how many tress are protected, the minimizing in the number of trucks carrying waste to dumps and the reduction in landfills. For energy, air discharges deliver the best effect, relating energy consumption to power plants and their emissions from burning fuel. All communications for environment and energy conservation have to be built on both a message and the medium. The message is what you are trying to persuade your occupiers to do, intended in a way to notice the listener. The medium is an instrument you use to transmit the message, whether it be a newsletter, meeting, posters or other methods. Use the mediums that function for your organization and the target audience. These include newsletter, memos, email notices, posters, information sessions and open houses. The message should be repeated frequently. In commercialising, repetition of the same message is considered significant to obtain the message across- utilize similar concept and deliver multiple communications.

Communication about environmental resourcefulness, whether for profitable goods or corporate facilities, must be focused at two different key targets, with different messages. The foremost target should be decision makers, those who approve or support your initiatives. For commercial buildings, this is the key tenant representative, preferably the ones accountable for the rent they pay. In corporate facilities, this audience is the executive and departments managers. Through collaborating to these people first, you can get buy-in to the definite application, both monetarily and operationally. With implementation supported from above, you can turn your attention to the occupants and design your communication to get their cooperation.

Decision Makers

Implementation may necessitate venture and will certainly entail buy-in from the inhabitants to get their submissive or active involvement, so you need support from the decision makers before you proceed. The decision makers may be within your own organization. Either way, their provision advances the probability of success for your creativity. Focus the communication with this group on the benefits they care about. Practically positively this will be financial, which can be driven from energy conservation, however it may also be the capability to present the organization and as a leader in environmental initiatives. You have to know the audience and highlight the aspects that will be most essential to them. If the decision depends on the payback calculations, put them front and center; however never eliminate the other benefits, delivering the icing on the cake. Formal communications are typically essential, including a business case and presentation. Don't rely just on formal channels though. Allot a time to talk to one-on one with the decision makers in an informal setting first. You can gain buy-in and address any issues they may raise before the formal presentation. Once you get buy-in, be sure to communicate your progress and if possible, send them duplicates of any communication you use with the tenants. If there is problem regarding with the implementation or acceptance arise, communicate your solution to them first, before anyone else can raise the issue. As the decision makers can help you with resistance to your initiative, it is essential to maintain communications and the relationship so you can use their influence effectively.

Communicating successfully with tenants when implementing initiatives should be done in all phases of your initiatives should be done in all phases of your initiative, not just during implementation. The common phases include Planning and Design, Implementation and Steady-state. Each requires a different communication devices.

Different Communication Mechanism

If you are equipped with provision from the decision makers, you may be interested to force implementation on the tenants. However reception of the inventiveness will be short lived, and you will rapidly see a deprivation in their performance. The finest way to utilize senior level support is at initiation point and to build buy-in directly through communication, encouragement and good application. Bear in mind that tenants will have a various perspective. Every individual is entitled to his/her own perspective. Energy conservation to save money isn't very important to them, as the savings are for the company. Focus on how it might impacted them, like as environmental issues, smog and clean air, occupant health, protection of hazardous and perilous materials out of the local landfill, fewer trees cut down, less water consumed and many more.

Planning and Design

When you have the ability and support, the first step is to communicate to the tenants stage, while doing nothing stage in order to construct the first step is to communicate to the tenants during the planning stage in order to build buy-accepted by tenants. Communications, in some circumstances, throughout this stage will need to go through or department representatives, but if it is possible, go directly to the tenant to confirm the message gets through. At this period, using a newsletter-even a single page format- is an efficient instrument for communicating mostly, but personal letters heading for key managers, union representatives, committees or alike stakeholders should be part of your communication plan. Open houses or data conferences should also be utilized for the tenants. Both supporters and detractors will be attracted with an open invitation. Identifying these people early will be a great help for testing the mechanism you have employed. The messages will be conveyed by the supporters and the detractors or critics will serve as the litmus test that will give the validity of your applications, and may provide a possibility to identify the best way to obtain buy-in. You can plan and set up a committee to review and suggest the implementation instrument, including communication and education equipment you need for the next level. The results must be communicated specifically when the implementation designed will be changed depending on the tenant's feedback- this will be a good way to show that you are listening to their concerns.

Implementation

Effective building of a new environment and energy conservation efforts such as a new recycling process or a lighting control methods relies on effective communication as a better plan. Various communication tools are present and are waiting for you to utilize, the more communication tools you practice, the better. Just continue with the newsletter and memos, aided with electronic mails, tent card, instruction cards and posters. If the effort is quite complex, such as a new reprocessing scheme that necessitates the resident to separate their garbage, simple instructions, graphics and other means that are readily accessed and referenced are critical. Information conferences and even presentations with usual garbage items will enhance the tenant's ability to comprehend the system. Applications such as telephone based lighting controls, instruction cards or stickers can be attached on phones to confirm the occupants who are actually utilize the system when they need it. Posters or floor plans in the elevator lobby with instructions are also helpful.

Steady state

Once the efforts are implemented, you have to continue communicating in order to sustain interest and participation. This communication should enclosed how successful the programs are and the explanations, if it is necessary. Like for an example, if you found out that specific recyclables are still generally not sorted out correctly, re-communicate about those issues particularly. Do not take the new occupants or employees for granted, they should be given appropriate and friendly communication as they arrive. Communicating the accomplishment of the program should give emphasis on the original messages that are important to the dwellers, strengthening the issue and congratulating the tenants for making a difference. To sustain attentiveness, try to implement some competitions within the building or between buildings, as they use waste diversion or proper segregation of garbage and consumption reductions as the strategy to be employed.

COMMUNICATE ENERGY EFFICIENCY

Figure 10.11: Plan and communicate energy efficiency.

Worthy communication increases consciousness of energy management and encourages employees to get on board and participate by doing their part. It also offers affirmative news to your company's stakeholders and can draw interest from senior executive assistance.

Many people are not skilled communicators, but with a little effort and few sacrifices, you can make an effective communication strategy and techniques. You can start designing the outline of annual energy program communication plan. Here is the list of steps you may follow.

Suggestion: Utilize your energy team to help carry the load. Make sure that the members of the corporate communications or marketing department is on the team.

STEPS TO FOLLOW IN BUILDING ENERGY PROGRAM COMMUNICATIONS PLAN

1. Define the goal

The plan must be composed of objective if it is going to be successful. Like for an instance, if you plan to increase the awareness about a particular energy system initiative or accomplishment or you are trying to influence a specific behavioural change, such as turning off equipment to save energy consumption, you may employ the work sheets that ENERGY STAR offers. The worksheet will help you in defining your goals.

2. Identify the target audience

Identify those people whom you need to approach. If you are targeting internal audiences such as employees or management design a seminar or symposium that is suited for them. If you plan to get involved with the external and public groups, such as customers, shareholders or the media, you may employ a seminar or training and other workshops that may help them to learn and for you to achieve your goals. You may invite speakers and facilitators from the government for such agency responsible for energy and waste management.

3. Tailor your messages

Plan and organize the key messages that you need to communicate and alter them to fit to your target audience. There are many ways hoe you may best communicate with your target audience. There are many videos that you may download and you may show them through seminars and symposium.

4. Select the best method to reach your target audience

In order to communicate effectively, identify the best method to present your message. You may ask the Human Resource Office and communications departments for the possible method that they may suggest to be the most efficient methods in communicating to various audiences. Through external communications, your organization may also have specific requirements and procedures that you must follow.

5. Schedule each activity

Identify the schedule on how often that communication may be needed, what resources are needed and who will be accountable. Finally, consider how you will may measure the impact of your communication efforts.

REFERENCES

1. Abramowitz, M. and Stegun, I. A. (Eds.). "Wave Equation in Prolate and Oblate Spheroidal Coordinates." §21.5 in Handbook of Mathematical Functions with Formulas, Graphs, and Mathematical Tables, 9th printing. New York: Dover, pp. 752-753, 1972.
2. Australian Science Education Project (ASEP).1974. Energy and Change. Australia: ASEP.
3. Baez, Albert V. 1967. The New College Physics, A Spiral Approach. San Francisco: W. H. Freeman and Company
4. Blackwood, D.H., Kelly, W. C., and Bell, R. M. 1963. General Physics, 3rd Ed. New York-London: John Wiley and Sons, Inc.
5. Blancard, C. H., et. Al. 1969. Introduction to Modern Physics. Englewood Cliffs: Prentice –Hall, Inc.
6. Brandwein, P. F., Stollberg, R., and Burnett, R. W. 1968. Energy: Its Forms and Changes. New York: Harcourt, Brace and World, Inc.
7. C.M. Lederer; J.M. Hollander; I. Perlman (1968). Table of Isotopes

(6th ed.). New York: John Wiley & Sons.

8. Collier's Encyclopedia. 1987. New York: MacMillan Educational Company.
9. Greenberg, Leonard H. 1978. Physics with Modern Applications. Philadelphia: W. B. Saunders Company.
10. Jardine, Jim, et.al. Physics through Applications.1998. Oxford: Oxford University Press.
11. Jennings, T. Sounds. Oxford New York, pp.7-10, 1984
12. J. Shapiro (1997). Radiation Protection: A Guide for Scientists and Physicians (3rd Edition)
13. Makhijani et al. (1995). Science for Democratic Action, vol. 6no. 2. Cambridge: MIT Press
14. Morse, P. M. and Feshbach, H. Methods of Theoretical Physics, Part I. New York: McGraw-Hill, pp. 124-125 and 271, 1953.
15. PLON. 1986. Ionizing Radiation. Netherland
16. Zwillinger, D. (Ed.). CRC Standard Mathematical Tables and Formulae. Boca Raton, FL: CRC Press, p. 417, 1995.
17. Zwillinger, D. Handbook of Differential Equations, 3rd ed. Boston, MA: Academic Press, p. 130, 1997.

Web Links

18. https://www.physicsclassroom.com
19. https://en.wikipedia.org/wiki/Wave
20. http://mathworld.wolfram.com/WaveEquation1-Dimensional.html
21. https://www.google.com.ph/webhp?sourceid=navclient&ie=UTF-8&gws_rd=ssl#q=traveling+waves
22. https://www.google.com.ph
23. https://en.wikipedia.org/wiki/Wireless
24. https://www.elprocus.com/types-of-wireless-communication-applications/
25. https://www.efxkits.us/different-types-of-wireless-communication-technologies/
26. https://www.physicsclassroom.com
27. https://en.wikipedia.org/wiki/Wave

28. http://mathworld.wolfram.com/WaveEquation1-Dimensional.html
29. https://www.google.com.ph/webhp?sourceid=navclient&ie=UTF-8&gws_rd=ssl#q=traveling+waves

INDEX

S

T

U

V

W